RIVER PLANET

RIVERS FROM DEEP TIME TO THE MODERN CRISIS

Also published by Dunedin Academic Press

Introducing Sea Level Change Dawson (2018)
Mountains, the origins of the Earth's mountain systems Park (2018)
Farewell, King Coal, from industrial triumph to climatic disaster Seaton (2018)
Introducing Hydrogeology Robins (2019)
Scotland's Mountain Landscapes, a geomorphological perspective Ballantyne (2019)
Measures for Measure, Geology and the industrial revolution Leeder (2020)

RIVER PLANET

RIVERS FROM DEEP TIME TO THE MODERN CRISIS

MARTIN GIBLING

DUNEDIN

EDINBURGH ◆ LONDON

Published by
Dunedin Academic Press Ltd
Main Office: Hudson House, 8 Albany Street, Edinburgh EH1 3QB, Scotland
London Office: 352 Cromwell Tower, Barbican, London EC2Y 8NB

www.dunedinacademicpress.co.uk

ISBNs
9781780460994 (Hardback)
9781780466583 (ePub)
9781780466590 (Kindle)
9781780466606 (PDF)

British Library Cataloguing in Publication data
A catalogue record for this book is available from the British Library

Typeset by Kerrie Moncur Design & Typesetting
kmdesigntypesetting@outlook.com

CONTENTS

SOURCED ILLUSTRATIONS

of the *Royal Society of Canada, Series 3* **50**, 3–12.
(2) Bornhold, B.D., Finlayson, N.M. and Monahan, D. (1976) Submerged drainage patterns in Barrow Strait, Canadian Arctic. *Canadian Journal of Earth Sciences* **13**, 305–311.

Figure 13.4: Alqahtani, F.A., Jackson, C.A.-L., Johnson, H.D. and Som, M.R.B. (2017) Controls on the geometry and evolution of humid-tropical fluvial systems: Insights from 3D seismic geomorphological analysis of the Malay Basin, Sunda Shelf, Southeast Asia. *Journal of Sedimentary Research* **87**, 17–40.

Figure 13.6: Gupta, S., Collier, J.S., Palmer-Felgate, A. and Potter, G. (2007) Catastrophic flooding origin of shelf valley systems in the English Channel. *Nature* **448**, 342–346.

Figure 13.7: (1) Bridgland, D.R. (2003) The evolution of the River Medway, SE England, in the context of Quaternary palaeoclimate and the Palaeolithic occupation of NW Europe. *Proceedings of the Geologists' Association* **114**, 23–48. (2) Gibbard, P.L. (2007) Europe cut adrift. *Nature* **448**, 259–260.

Figure 14.2: Gibling, M.R. (2018) River systems and the Anthropocene: A late Pleistocene and Holocene timeline for human influence. *Quaternary* **1**, doi:10.3390/quat1030021.

Figure 14.5: Gibling, M.R. (2018) River systems and the Anthropocene: A late Pleistocene and Holocene timeline for human influence. *Quaternary* **1**, doi:10.3390/quat1030021.

Figure 14.8: Lightfoot, D.R. (1996) Syrian qanat Romani: history, ecology, abandonment. *Journal of Arid Environments* **33**, 321–336.

Figure 16.5: Chu, Z. (2014) The dramatic changes and anthropogenic causes of erosion and deposition in the lower Yellow (Huanghe) River since 1952. *Geomorphology* **216**, 171–179.

Figure 17.3: United States Geological Survey.

Figure 18.2: Photo by Per-Olow Anderson.

Figure 19.2: Smith, N.D., Morozova, G.S. and Gibling, M.R. (2014) Channel enlargement by avulsion-induced sediment starvation in the Saskatchewan River. *Geology* **42**, 355–358.

Figure 20.4: Contreras, S., Santoni, C.S. and Jobbagy, E.G. (2012) Abrupt watercourse formation in a semiarid sedimentary landscape of central Argentina: the roles of forest clearing, rainfall variability and seismic activity. *Ecohydrology* **6**, 794–805.

Figure 21.2: Barton, N. (1962) *The Lost Rivers of London*. London: Historical Publications.

All other illustrations are by the author.

ACKNOWLEDGEMENTS

I am more deeply indebted than I can express to many friends and colleagues for their assistance and friendship. John Burgess, Elisabeth Kosters, and Marcos Zentilli generously read the whole manuscript and provided vital overviews. Steven Andrews, David Bridgland, Carol Bruneau, Gary Carriere, Dave Craw, Bob Diffendal, Bill DiMichele, Ann Emery, Ellen Gibling, Felicity Gibling, John Gosse, Sanjeev Gupta, Zehui Huang, Alessandro Ielpi, Dale Leckie, Galina Morozova, Peta Mudie, Mike Rygel, Myrtle Shock, Norm Smith, Sampat Tandon, Ronald Van Balen, Steve Vincent, Gary Weissmann, and Rebecca Williams commented in depth on individual chapters. I would like to thank those who provided photographs, as recorded in the credits, and to Gary Weissmann for generously working with satellite imagery for illustration.

Many friends alerted me to river literature, historical accounts, and current events. They include Kent Arnold who took me to the site of the Lawn Lake dam failure; Gail Ashley who shared information about early hominins in East Africa; John Calder and Howard Falcon-Lang for their biographic knowledge of early geologists; Gary Carriere and friends at Cumberland House who introduced me to the Saskatchewan River Delta; Don Cotton who showed me river histories in Ireland; Bob Dott for taking me to Aldo Leopold's Shack; Rebecca Jamieson for advice about zircons; Rama Mchomvu for guiding me to Olduvai Gorge and the Serengeti Plains; the First World War group of Halifax, organized by Mark Sadler and Susan Rahey; Vincenzo Pascucci for introducing me to the geology of Italy; Elaine Russell and students at Hammonds Plains Consolidated School for welcoming me to their classroom; S. Swaminathan for an introduction to the Rigveda; Don Wightman for an artistic insight into the Universe; Margie Wightman for information on family history on the North China Plain; and Rebecca Williams for showing me the geology and landscapes of Mars from her basement. Alan Edwards of Tiverton Grammar School introduced me to geology, a lifelong gift.

Colleagues and students at many institutions made an invaluable contribution to my understanding of rivers. I think especially of Harold Reading at Oxford University; Owen Dixon and Brian Rust at University of Ottawa; colleagues and students at Chiang Mai University in Thailand; Sampat Tandon, Rajiv Sinha and students at Delhi University and Indian Institute of Technology at Kanpur; Brian Jones, Jerry Maroulis, Gerald Nanson and colleagues at University of Wollongong; students at Lanzhou University who knew the poems of Li Bai by heart; and many colleagues and students at Dalhousie University who shared field excursions with me over the years, passing on their insight and observations. I am grateful to the staff at the Killam Library at Dalhousie University for their help over many years in accessing the literature. Many friends and members of the public gave me enthusiastic support.

I am grateful to Göran Molin for translating the obituary of Assar Hadding, and to colleagues at the University of Lund who identified Hadding's original microscope sections. Nancy Zhang kindly translated part of the text of the Sanmenxia Cantata. Fieldwork over the years was largely funded by the Natural Sciences and Engineering Research Council of Canada.

I am indebted to Anthony Kinahan and all working with him for Dunedin Academic Press for seeing this book through to publication. I am grateful to Meredith Sadler for producing superbly drafted diagrams from my clunky originals.

Finally, I owe a huge debt of gratitude to Maureen, Ellen and David for sharing with me so many experiences that contributed to the shaping of this book, as well as their unstinting support of the fieldwork and writing. My sisters Rosalind, Stella, and Felicity provided strong encouragement. And my parents Robin and Heather Gibling and Uncle David passed on to me their deep love for the Earth and a sense of wonder about all its creatures.

PROLOGUE

Rivers were close at hand when I grew up in the town of Tiverton in southern England. Stone Age people had lived by the Exe and Lowman rivers, and Civil War cavalry had skirmished at Tumbling Fields, unhorsed on the tussocky water meadow. Water power from the rivers had attracted a textile factory, and every spring in my childhood, excitable cattle leapt a fence into the factory channel and were wrestled out by the farmer as the factory workers cheered. Children celebrated the medieval gift of a small canal by whacking the water with sticks every seven years.

At the age of eight with a tyrant as a teacher, I trudged to school every morning in a sorry state of mind, crossing the River Lowman by an uninspiring statue of King Edward VII and an iron foundry that emitted pillars of flame. But below King Edward, fish and tadpoles could be collected from the river, as well as caddis-fly larvae encased in twigs, and once an eel-like lamprey came writhing up the channel and latched onto my rubber boots. A friend and I punted up and down on a raft made from a wooden pallet and empty oil drums purloined from the nearby brewery.

Summoned from school to deal with an imminent floodwave on the River Exe, we members of the Flood Squad filled sandbags through the night and stacked them at a low point in the bank as the swirling waters rose steadily higher. At dawn the flood peaked below danger level and we had saved the town and—more significantly—got the next day off school. In this respect we were successors to students at nearby Blundell's School, as recounted by R.D. Blackmore in his novel *Lorna Doone*. When the rising Lowman lapped the initials of the founder, Peter Blundell, laid out in white pebbles by the school gate, any boy had the right to rush into the school and scream 'P.B.'. At this news, the boys flung caps and books in the air as they rushed out, leaving the masters to their pipes and foreign cordials.[1]

Once I cycled to the source of the River Barle, a tributary of the Exe, at Pinkworthy Pond where springs welled from the blanket bogs of Exmoor. The pond, its purpose unknown, drained into the river through a rock-cut tunnel: was it for irrigation, for driving agricultural machinery, or to improve the scenery? A century earlier, love-sick Farmer Gammon had tied his pony to a tree and drowned himself in the pond. When a candle floating on a loaf of bread failed to rest over the body, villagers drained the pond and found the unfortunate farmer. For a long while I sat by the sombre lake, wondering about his suicide but exhilarated at reaching the river's source. The drone of a plane was the only sound, and I felt intensely alone at the top of the world.

It did not occur to me that rivers might have a history as intimate as any family tree, merging, diverging, or dying out. My eyes were opened to river history during a school excursion to the cliffs of Wales, where layers of Old Red Sandstone—former river sand turned to rock—rested on layers with marine fossils. In 1914 survey geologist Thomas Cantrill had carved an arrow on the cliff to mark where the long-extinct rivers had advanced across the sea floor, defining a body of river rock that would prove to be of great significance for geology and botany (see Chapter 3). It was a revelation to me that these rock strata represented the Earth's ancient landscapes. In the words of Welsh miner and poet Huw Menai, 'geology is the infinite biography of God'.[2]

During a university mapping project, I studied the Old Red Sandstone on the Shetland Islands north of the Scottish mainland, camped by a graveyard where a derelict ice-cream van, its gaping roof plugged with a car hood, sheltered me from the Atlantic storms. I found the fossils of river fish and ripple marks on the sandstone beds where ancient rivers had sculpted the

Figure i.i Dawn on the Tana River, Norway.

sand. Outdoor geology drew me in. In his biography of Alfred Wegener, Mott Greene has noted that humanity is divided into those who cannot live without feeling the wind in their faces and those who can.[3] I discovered that I could not.

The first sandy river that I encountered was the Tana, which crosses the Norwegian tundra to the Arctic Ocean (Fig. i.i). Shrouded by a dawn mist in the short summer months, the river wound its way through a mosaic of sand bars with ripples like those of the Old Red Sandstone. At the river mouth large cod seemed almost to leap onto the fishing lines. The second river was the San Juan, a sandy channel flowing through the deserts of Utah.

For a while I contemplated a career in marine geology and joined a winter oceanographic cruise off eastern Canada. While others made scientific discoveries, I lay seasick in my bunk as the pack ice hurtled past a porthole where the horizon tilted alarmingly. Back on terra firma, I concluded that my manifest destiny lay with rivers.

Some years ago, I led an excursion to the Joggins Fossil Cliffs in Canada (see Chapter 3). A score of fossilized trees stood upright in the cliffs, preserved in Carboniferous river strata 300 million years old. The enthusiastic and perceptive participants were of all ages and from all walks of life. We discussed how river floods had swept through the coal swamps and entombed the trees.

'I wish I knew more about this', said one participant suddenly. I replied that I had considered writing a book about rivers and their geological history. 'I would read a book like that', he replied.

This, I hope, is such a book—a history of the River Planet seen through the life of its rivers. The account begins with the first drops of rain more than four billion years ago and continues with the long-extinct rivers that crossed supercontinents until vegetation covered the land and new river forms emerged. Most of our modern rivers came into being when the supercontinent of Pangea broke apart about 200 million years ago, setting the continents adrift. There are rivers created and destroyed by Ice Ages, most recently over the last few million years, and there are rivers altered beyond recognition by humans over the past centuries.

The book is a celebration of people who were fascinated by the Earth and its history. Stonemasons, bakers, janitors, physicians, woodcutters, farmers, and landed gentry—all represented in these pages—puzzled over the rocks and fossils, recording vivid observations and new thoughts. There are scientists who unravelled the geological ancestry of rivers, and there are poets and writers who understood rivers in ways beyond science, as well as activists who fight for the rivers they know. But no one knows rivers more completely than the Earth's indigenous peoples, who have always been part of nature with all its rights and responsibilities, privileges and fears. For them, rivers have been a source of spiritual insight, sacred mystery, and delight.

As I embarked on this river journey, I let curiosity rule the day. River history soon became intertwined with my own history, and I often turned off the highways of science into untrodden paths. If there is a plot, it resides in the whimsical movement of flowing water. As I learned more about rivers and how we are destroying them, I became less impartial and more passionate.

This book is an unauthorized biography of rivers. Listening to the river's song, perhaps I may dare to be the river's voice for this short time. Eventually I will have to let the rivers go, reflecting with the dying Victorian naturalist Richard Jefferies, 'I wonder to myself how they can all get on without me—how they manage, bird and flower, without me to keep the calendar for them. For I noted it so carefully and lovingly day by day'.[4] Until then, however, I may follow in the footsteps of environmentalist John Muir. Asked whether giant redwoods would make good furniture, Muir replied 'Would you murder your own children?'[5] For let us be clear: we have squandered four billion years of river history in less than three centuries, and rivers are an endangered species.

PART 1: RIVERS IN DEEP TIME

CHAPTER 1

RIVERS AND GEOLOGICAL TIME

The Earth has a history

How can we know about rivers that became extinct millions of years ago? The flowing water has gone. The channels and river plains have gone. Even the continents have gone. Only the river sediments of gravel, sand and mud remain, turned into rock as conglomerate, sandstone and mudstone through geological time.

Prior to the observant eye of Leonardo da Vinci in the fifteenth century, the connection between soft sediment and solid rock was obscure. The rocks were a pre-existing backdrop for the human drama. People had a history, the Earth did not. Villagers around the world equated fossils with fallen angels, witches, and fairies.[1] Black curved oyster shells were the devil's toenails, and the coin-shaped fossils of protozoans in the slabs of the pyramids were lentils from the lunches of the slaves. Belemnites, the pointed shells of fossil squid, were thunderbolts: how had they got into the rocks, people wondered?

Much as kidney stones grow within the human body, fossils were widely supposed to have grown in the rocks, perhaps under the mystical influence of the stars.[2] Far ahead of his time among Europeans, Leonardo observed that fossil shells with growth lines resembled the clams on a modern beach, implying that they had once been living creatures.[3]

After Leonardo came a succession of observers, many of whom were rewarded by posterity with grandiose epithets. There was Nicolaus Steno, the seventeenth-century Danish anatomist who first understood that the upward succession of strata marked the passage of time. There was James Hutton, the Father of Modern Geology, who identified unconformities in the strata and grasped the immensity of geological time. William Smith, the Father of Stratigraphy, was a blacksmith's son who played marbles with spherical fossils and produced the first comprehensive geological map. There was Mary Anning, the remarkable Finder of Fossils who discovered large Jurassic reptiles, and Etheldred Benett, scientist and artist of fossils and considered the First Lady Geologist. Other famous contributors included Charles Lyell, the Father of Uniformitarianism, Charles Darwin, the Father of Evolution, and Louis Agassiz, the Father of Glaciology. There was Eduard Suess, the unraveller of mountain belts and the Father of Urban Geology, who designed an aqueduct from the Alps to save Vienna from water-borne epidemics. And there was the eccentric Marie Stopes, the mother of almost everything from palaeobotany to womens' rights, family planning, and sensual literature.

Unconformities and the discovery of Deep Time

On a spring day in 1788, James Hutton and his friends Sir James Hall and John Playfair sailed along the Scottish coast. At Siccar Point they discovered geological time.

Hutton had studied in Edinburgh, understanding from Newton's Laws that the Earth behaved like a gigantic machine.[4] Apprenticed in a law office, he entertained his fellow clerks with chemical experiments and later obtained an income by extracting ammonium chloride from soot. Growing interested in geology, he concluded that, as far as observation could discern, the world had neither a beginning nor an end.

Hall and Playfair were impressed by Hutton's insistence that Earth history did not require catastrophes, just large amounts of time. And time, he insisted, was available. In the cliffs at Siccar Point they found evidence for his assertion. Steeply dipping beds of grey sandstone were cut by a jagged surface that was covered by gently dipping beds of red sandstone and conglomerate (Fig. 1.1). Near the point the waves had removed the softer red rocks, and the men walked across the exposed

Figure 1.1 Hutton's Unconformity at Siccar Point, Scotland, discovered in 1788.

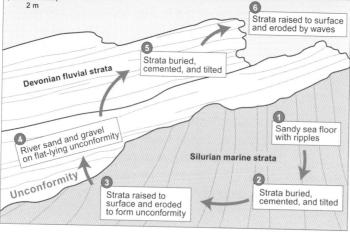

surface of an *unconformity* that separated the two groups of strata. Hutton deduced that the unconformity was the eroded rocky surface of a former upland carved by rivers.

But Hutton deduced more than this. The grey sandstone had rippled surfaces and must once have been soft sand moulded by waves and currents, like those on any beach. They must have been deeply buried and turned to rock as the sediment grains were pressed together and crystals precipitated in the interstices. Earth movements had tilted the rocks and brought them to the surface, where ancient rivers eroded and buried them under sand and gravel, later turned to rock, as the land subsided. It must all have taken a very long time.

The rock record was logical, not magical. The present with its largely humdrum physical events was the key to understanding the past—the principle of *uniformitarianism* that Charles Lyell later promoted. Hutton demonstrated the immensity of geological time, which allowed this principle to operate. John Playfair, who had a way with words, brought *deep time* to the public in his biographical account of Hutton:

The mind seemed to grow giddy by looking so far into the abyss of time; and while we listened with earnestness and admiration to the philosopher who was now unfolding to us the order and series of these wonderful events, we became sensible how much farther reason may sometimes go than imagination can venture to follow.[5]

Over the following century geologists observed that certain strata contain distinctive fossils. While working as a canal surveyor, William Smith discovered that the fossils could be used to order the strata from older to younger, and in 1815 he constructed a geological map that followed the strata across England and Wales.[6] The concepts were passed down to newcomers. The elderly James Hall took a young Charles Lyell to Siccar Point, an experience that contributed to Lyell's Principles of Geology.[7] Charles Darwin heard lecturers in Edinburgh variously laud and lambast Hutton, and he carried the first volume of Lyell's book onboard HMS *Beagle*, a key factor in his formulation of biological evolution.

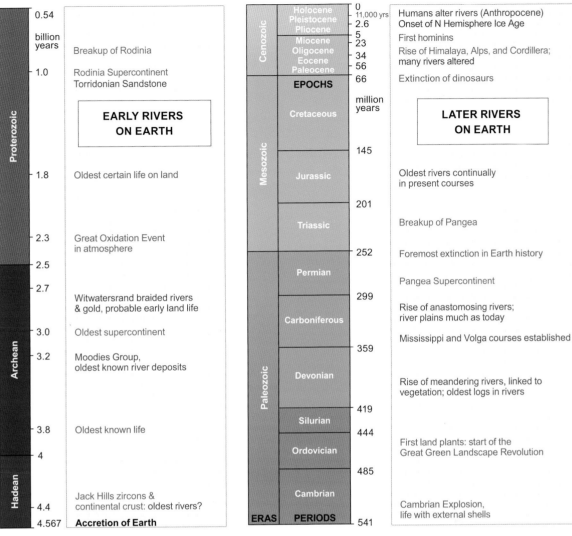

Figure 1.2 Geological timeline for the evolution of the Earth's rivers and other key events. Age dates from Ogg et al., 2016.

Fossil experts began to construct a relative timescale for Earth history (Fig. 1.2). Preservable shells evolved during the Cambrian Explosion, and notable appearances and extinctions of fossils allowed geologists to distinguish Paleozoic, Mesozoic and Cenozoic eras (ancient, middle, and new life), Cambrian to Cretaceous periods, and Paleocene to Holocene epochs. The Ice Age in the Northern Hemisphere effectively spanned the Pleistocene epoch (see Chapter 11).

But no one knew how old the Earth really was, and geologists often recklessly assumed almost infinite amounts of time.[8] Finally, at a 1904 meeting, nuclear physicist Ernest Rutherford suggested that newly discovered radioactivity could be used as a 'clock' to date rocks by applying rates of radioactive decay. In the audience was an aging Lord Kelvin who had used a cooling model to estimate that the Earth was 20 million years old. Kelvin slept through the meeting but awoke for the important moments. In 1907 Bertram Boltwood worked out the first numerical dates for rocks and minerals, including a date of 2.2 billion years for a sample of the mineral thorianite from Sri Lanka. Other rocks too old to contain shelled fossils yielded dates that were used to define the Archean and Proterozoic eons. The boundaries between the younger geological periods and epochs were dated using lava and ash interbedded with fossiliferous strata. During the latter part of the twentieth century, the development of radiocarbon dating and optical luminescence dating, which estimates the time since a sediment was buried away from sunlight, provided dates for sediments hundreds of years to a few tens of thousands of years old, linking geology to human history and archaeology in the Pleistocene and Holocene. And at the youngest end of the scale is the Anthropocene, broadly defined as the period of human impact on the Earth, and likely soon to be defined more precisely by the 'Great Acceleration' in about 1950.

Cross-beds and ancient river flow

Hutton had identified ancient river landscapes, but Henry Clifton Sorby in the mid-1800s began to explore the sediment and flow of ancient rivers. Sorby came from a family of Sheffield cutlery manufacturers in northern England, and able to pursue his interests through private means, he became fascinated with geology through the writings of John Playfair.[9] One day, sheltering from a rainstorm in a quarry of Carboniferous sandstone, Sorby was struck by the resemblance between ripples preserved in the sandstone overhead and those forming in the nearby river sand. Experimenting with a brook where he could control the flow and supply of sand, he calculated the water depth and speed under which ripples and larger subaqueous dunes were formed. Observing inclined layers within the nearby beds of sandstone, he deduced in an 1859 paper that the sand had slid downflow over a steep dune face to form 'drift-bedding' or cross-beds (Fig. 1.3), and he measured the dip directions of 20,000 cross-beds around Sheffield to work out which way the Carboniferous rivers had flowed.

Other researchers noted boulders that dipped upstream, stacked against each other like terracotta roof tiles by the force of floodwater. Termed *imbrication* from the Latin *imbrex* for a tile, the tilted boulders provided a clue to flow directions in conglomerates (Fig. 1.3).

Nearly a century later, Sorby's contributions were extended by Daryl Simons, who grew up as an orphan with his tough older brother on a Utah farm.[10] While working as a carpenter at a college construction site, Simons continued work after other carpenters downed tools following arguments with an irascible Dean. Asked why he hadn't quit, Simons replied that the Dean didn't

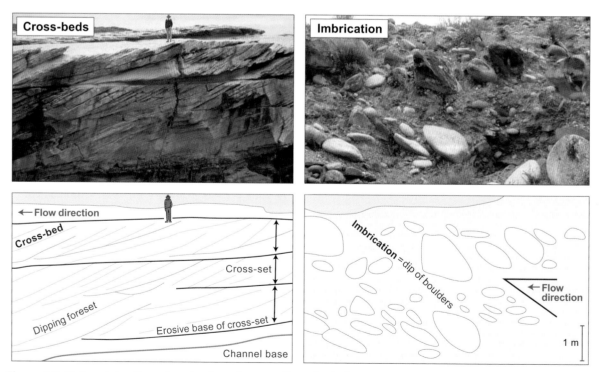

Figure 1.3 Determining flow direction in rock strata. Cross-beds from the Triassic Hawkesbury Sandstone, Australia. Imbrication of boulders in the Cenozoic terrace deposits of the Yellow River, China.

know his older brother. Impressed by his persistence, the Dean encouraged him to study engineering, and after war service in Patton's army, Simons experimented with channel flow using modern rivers and an artificial channel 60 m long called the 'Grand Canyon'. By the 1960s, his group had extended the principles of sand transport and deposition, later applying this knowledge to flood control in the Indian subcontinent and around the world. A generation of students recalled Simons' work from a movie in which surveyors sank up to their armpits in the Rio Grande.

Geographers had long studied surface drainage systems. A stream order of one had been assigned to the smallest, most elevated channels, progressing to higher orders as the channels picked up tributaries downflow. *Dendritic*, tree-like patterns of branching characterized flat-lying drainage basins with relatively uniform rock or sediment, whereas *rectangular* or *trellis-like* patterns characterized drainage basins with linear features such as faults (see front cover). With highly variable river flow and sediment transport, flood regimes were important geographic attributes. In alluvial channels, flow reaches the tops of the banks (bankfull level) every few years, spilling over onto the adjacent *floodplains* with which the channels are connected. In contrast, channels running through upland bedrock tend to occupy *valleys*, which are wider and deeper than the channels within them and incised too deeply for the flow to spill over the interfluves except during rare catastrophic events. Such channels are largely *disconnected* from their surroundings.

River *terraces* were first described during the Lewis and Clark expedition of 1803-6 in the midwestern United States,[11] and they provided important clues to river history and valley formation. Above the active channels were single or paired terraces of sand and gravel that marked former courses left stranded when the rivers cut down during periods of tectonic uplift. Terraces are common on rivers around the world and are prominent along the rivers of southern England and northern France, where the land has been rising slowly through time. Good examples are present along the Thames, where Trafalgar Square in the heart of London was built on a river terrace (Fig. 1.4).

Geography and geology had been widely regarded as distinct disciplines, but the studies of Simons and others began to link them more closely. Armed with a new understanding of flow dynamics, geologists began to examine the sediments of modern rivers in trenches and cores and using imaging methods such as ground-penetrating radar, applying their knowledge to the interpretation of ancient river strata, many of which formed important aquifers and reservoirs of oil and gas. On modern river plains, *meandering, braided,* and *anastomosing* rivers were important components in a spectrum of surface planforms, and these river styles could often be inferred for strata in the rock record (Fig. 1.5). Where rivers reached the ocean and fanned out across low-gradient deltas, *distributary channels* were prominent, as in the Lena Delta of Siberia, building out into the Arctic Ocean (Fig. 1.6).

Meandering rivers

Ask a child to draw a river and a wavy line will appear—the meandering river named for the Büyük Menderes

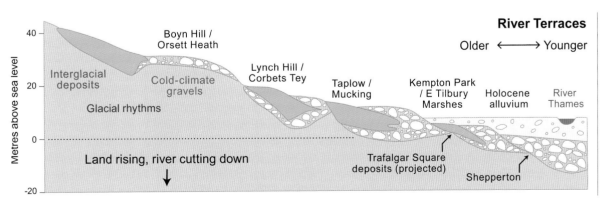

Figure 1.4 Terraces of the River Thames near London, UK. After Bridgland and Westaway, 2008.

Braided River

Meandering River

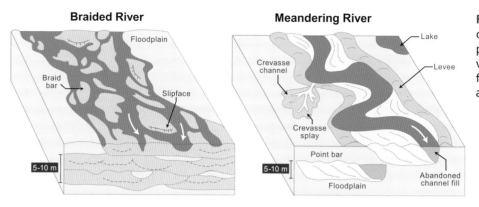

Anastomosing River

Valley System

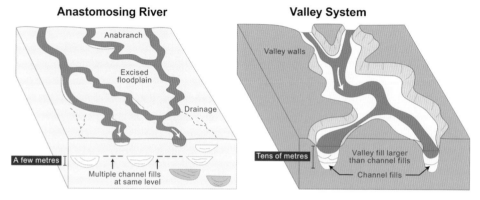

Figure 1.5 Illustration of common river planforms and valleys, with the front face showing their appearance in outcrop.

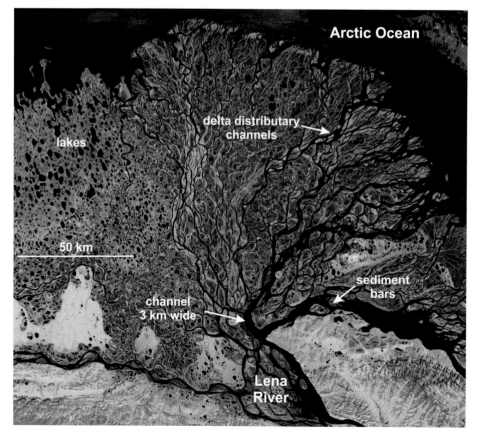

Figure 1.6 Lena River Delta, Russia, in false-colour image. Courtesy of USGS EROS Data Center Satellite Systems Branch, acquired by Landsat 7 Enhanced Thematic Mapper on 27 Feb., 2000.

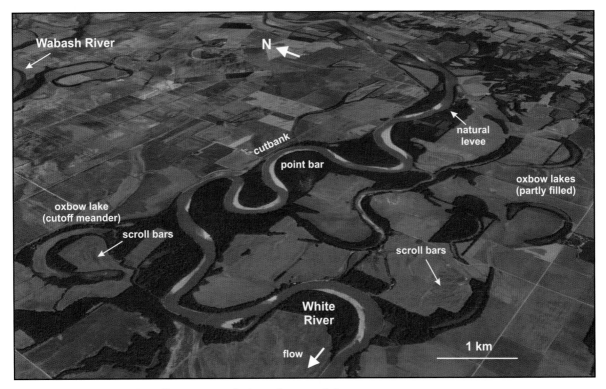

Figure 1.7 White River in Indiana, a meandering tributary of the Wabash River in the Mississippi catchment. Maps data: Google.

River in Turkey, the winding Maeander of Homer's *Iliad*. With its banks strengthened by roots and stiff mud, a meandering river is stable but not permanently fixed (Fig. 1.7). Turbulent flow hits the curve of the bend and erodes a *cutbank*, spiralling downstream towards the opposite bank where a sloping *point bar* of rippled sand builds up, with low curved ridges or *scroll bars* where sand gathers behind bushes. Floods lay down sand and mud as *natural levees* along the channel banks, bordering a muddy floodplain. Pulled by erosion on one side and pushed by sediment accumulation on the other side, the channel migrates slowly across the plain to form a sheet of sand several metres thick. The next bend downstream migrates in the opposite direction, exaggerating the winding form, and floods eventually carve *cutoffs* through the narrow meander necks. The former channel bend is left as an abandoned, curved water course, an *oxbow lake*, named for its similarity to the U-shaped collar placed around the neck of an ox. If this process continues, a wide swathe of cutoff, intersecting meanders develops—a *meander belt*. The point bars contain dipping layers of sand and mud that, along with scroll bars, help to identify meandering rivers in the rock record.

Researchers found it difficult to generate meandering rivers in experimental channels. Success came when they sprinkled the sand with rapidly germinating alfalfa, which was so tenacious that flows failed to dislodge the rooted plants, developing migrating bends. Aided by mud that strengthens the banks and clogs breaks, a stable meandering river can form.[12]

Our knowledge of meandering rivers stems from the classic work of Harold Fisk on the Mississippi. Fisk's 1944 maps show a complex mosaic of cut-off meander loops, point bars, and oxbow lakes in a meander-belt as much as 15 km wide (Fig. 1.8).[13] But in 1973, the US Army Corps of Engineers nearly 'lost' the Mississippi at the Old River Control Structure (Fig. 1.9). The river has a long and gentle course through New Orleans to the Gulf of Mexico, whereas the nearby Atchafalaya River is shorter and steeper—a more advantageous course that the Mississippi will exploit some day with a breakout or *avulsion*. The 1973 flood scoured deeply around the control structure, and the Corps just managed to

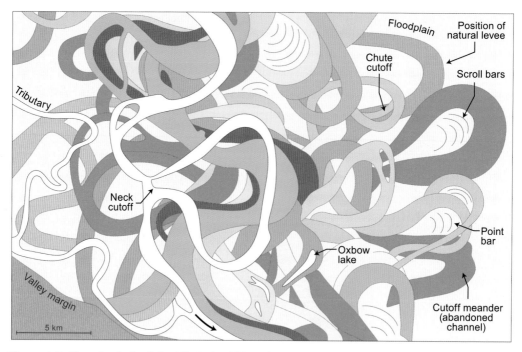

Figure 1.8 Meander-belt of the Mississippi River, formed by repeated meander cutoffs, near Natchez, Mississippi. After Fisk, 1944.

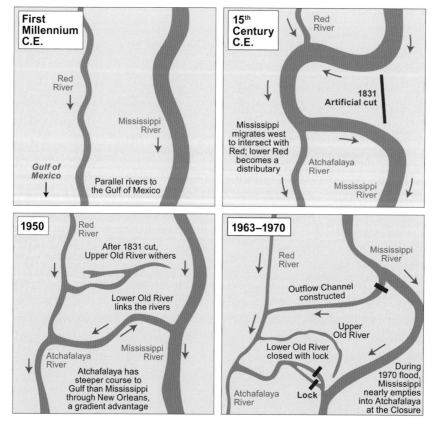

Figure 1.9 Evolution of the Mississippi, Red, and Atchafalaya rivers at Old River, Louisiana. After United States Army Corps of Engineers.

keep the Mississippi in its course through New Orleans by dumping rocks in the growing hole.[14] In the words of former steamboat pilot Mark Twain in *Life on the Mississippi*,

> One who knows the Mississippi will promptly aver—not aloud but to himself—that ten thousand River Commissions, with the mines of the world at their back, cannot tame the lawless stream, cannot curb it or confine it, cannot say to it, Go here, or Go there, and make it obey; cannot save a shore that it has sentenced; cannot bar its path with an obstruction which it will not tear down, dance over, and laugh at.[15]

Braided rivers

Braided rivers are the workhorse of the fluvial world, emerging from the mountains as steep channels with a multitude of sand and gravel bars. During floods only a single turbulent channel is visible, but as the flow wanes the bars break the surface and divide the flow into smaller channels like plaited braids of hair. Sand pushing over the bar slipfaces creates cross-beds like those that Sorby described, and pioneering plants colonize the loose sediment to form vegetated islands.

Among the most famous braided rivers is the Platte. In the 1840s, thousands of settlers trekked into the American West along the Oregon Trail. The journey took some six months if all went well, but often it did not, and one in ten travellers may have perished. The oxcarts required a gentle pass through the mountains with water and feed for the animals, and the settlers found such a route along the Platte River with its broad shallow flow (Fig. 1.10). The wagons toiled up the river valley past the dramatic spine of Chimney Rock that, unknown to them, exposed strata laid down by an ancestral Platte River. Where the wagons breasted the steep valley margins, deep ruts still record their passing.

But the pioneers did not warm to the Platte—wide but hopelessly shallow, murky yellow with sediment but not solid enough for ploughing. Enduring blizzards and droughts, those that settled along the seasonal

Figure 1.10 Platte River near Lincoln, Nebraska, a sandy braided river with braid bars (some partially vegetated) and multiple shifting channels of various sizes.

river were tough and humorous people. One drought-stricken farmer claimed to have seen four cottonwood trees chasing a single dog, and a dustbowl farmer is said to have fainted after being hit by a drop of water, only reviving when three buckets of sand were thrown over him.[16]

The Platte is famous among geologists for the pioneering studies of Norman Smith, who described the sand bars and channel fills that also identify braided rivers in the rock record.[17] The river is also famous for half a million sandhill cranes on migration, gliding down at dusk on stiff wings to roost on the sandbars.

Anastomosing rivers

My thermometer registers 43°C in the shade of a tree. But there is no shade in the nearby dry creek, where the four-wheel-drive is hopelessly bogged. The rainfall in this part of Queensland is less than 200 mm a year and mercifully none of it is due for months. We dig out the wheels and tie the winch to the only log within reach, which splinters under the strain. Leaning against a eucalyptus tree, I am startled by a feral pig awakened from sleep in the hollow trunk. Eventually a truck from a nearby oilfield pulls us out. I feel beyond rehydration.

These are some of the oddest rivers on Earth—the anastomosing Georgina, Diamantina, and Cooper Creek of Australia's Channel Country.[18] Each river may have as many as twenty channels but, unlike the shifting channels and bars of braided rivers, the channels run between stable islands of floodplain mud, endlessly anastomosing, branching and rejoining (Fig. 1.11). Crossing the world's driest continent, the rivers variously trickle or flood, and multiple channels may be the river's way of moving water across flat plains in a strategy of divide and conquer. The muddy plains are so tough that even a backhoe may not penetrate the mud, and mud plasters the channel banks as the floodwaters recede, progressively narrowing the channels. Suites of narrow channel fills encased in mudstone may represent anastomosing rivers in the rock record.

The Channel Country rivers drain more than a million square kilometres, brimming with water for a month a year on their way to Kati Thanda (Lake Eyre), which floods about every decade as millions of waterbirds congregate, mysteriously aware that the 'inland sea' has reformed. The floods may create a lake

the size of an American state, where only the topmost branches of the riverbank trees mark the channels.

Aboriginal peoples named the permanent waterholes on which their lives depended—Tooley Wooley, Chookoo, Tookabarnoo, hemmed in by sand dunes. The wealth of the land was theirs, the kangaroos and emus that left their tracks in the channels, and the crayfish that burrowed down to the water table far below the surface during the dry seasons. When the wind shifts the dunes, delicate stone arrow heads and scrapers reappear. Long before Europeans explored the inland rivers, the first inhabitants had mastered the harsh environment of the Channel Country.

Tracing ancient rivers in Nebraska

From the 1950s onwards, geologists began to identify braided, meandering, and anastomosing rivers in the rock record. One important body of braided-river strata was the Ogallala Group, dating back to the Miocene and named for a post office on the South Platte River. The Ogallala strata cover nearly half a million square kilometres from South Dakota to the Texas panhandle, where the High Plains slope southeast to the Gulf of Mexico. The modern rivers have cut through the Ogallala river strata, which are exposed in the valley walls. The strata were laid down on *megafans* and smaller *alluvial fans* (distributive fluvial systems), formed where the gradients of mountain rivers decreased as the channels entered the plains and lost transport capacity (Fig. 1.12). Fanning out across the basin in a similar manner to delta channels, the rivers periodically switched their courses during floods and combed across the fans in response to sediment buildup and small advantages in gradient. Along the Cordilleran mountain front to the west, the river exits—and thus the positions of megafans and alluvial fans—showed a regular spacing that, in modern uplands, is related to the width of the mountain belt from the 'ridgepole' of the range to the foothills, linked to the uplift and erosion history.[19] The Ogallala strata host the High Plains Aquifer, a vital but depleted source of groundwater for irrigation.

Across the Nebraska plains, field mapping of the Ogallala strata fell to geologist Bob Diffendal. With a hammer in a belt harness and a leather pouch with fieldbooks, maps and instruments, Bob set out to walk

Figure 1.11 The Diamantina River in the Channel Country of Queensland, Australia, an anastomosing river with multiple stable channels cut into tough floodplain muds.

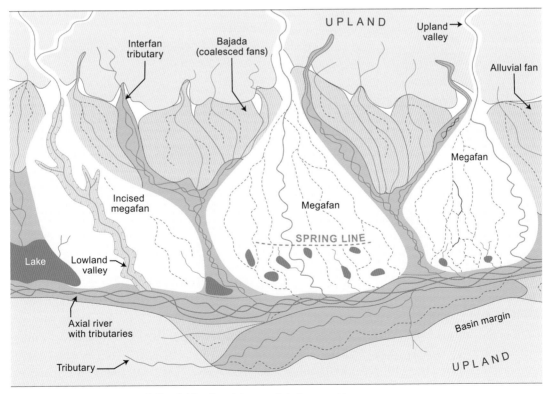

Figure 1.12 Illustration of alluvial basin with an axial river and transverse megafans and smaller alluvial fans building out from upland valleys. After Decelles and Cavazza 1999, and Weissmann et al., 2015.

Figure 1.13 Braided-river conglomerates and sandstones of the Ogallala Group of Miocene age, Nebraska.

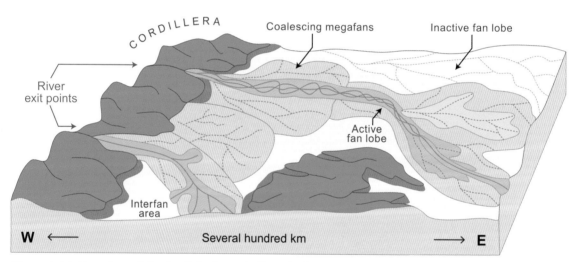

Figure 1.14 Megafans of the Ogallala Group in New Mexico and Texas, building out from exit points in the Cordillera. After Seni, 1980.

the ranges, holding his walking stick in front of him so that unseen rattlesnakes would strike the stick rather than his legs. And, indeed, the stick is darkened and mottled from the venom of many snakes. Bob obtained permission from every rancher on the Ogallala in the state, shooting the breeze among 1930s trucks. His only companions in this land of big skies and tumbleweeds were cattle, coyotes, and the sandhill cranes that flew north as the snow melted from the valleys.

The Ogallala marks the rise and fall of an ancient landscape.[20] Huge eruptions from volcanoes in the western mountains covered the plains with ash, and the Ogallala braided rivers carved valleys through the volcanic debris, filling their courses with cross-beds of sand and gravel (Fig. 1.13). Slump blocks tumbled from the valley walls and may have blocked the channels until the rivers broke through. Following his geological maps but mostly from memory, Bob can trace former valleys that were 50 m deep and nearly two kilometres wide, mapped so precisely that he can estimate their former slopes. The cross-beds show that the megafan rivers flowed southeast towards the Gulf of Mexico at a high landscape level, breaking out at times to generate new fan lobes (Fig. 1.14) until the High Plains were tectonically uplifted and the modern Platte cut through the older strata.

So abundant are the bones of prehistoric vertebrates in the Ogallala strata that many highway cuts yield rich finds. There are bones of rhinos, horses, and giant camels that lived and died along the streams. About 12 million years ago in the Miocene, a volcano from the Yellowstone area blew its top in an eruption larger than any historic blast, blanketing Nebraska with several metres of ash. At Ashfall Fossil Beds State Historical Park, the ash overwhelmed a herd of rhinos at a waterhole, along with horses, camels, turtles, a beaver, a constrictor snake, and cranes. Not even flight could save the birds.

CHAPTER 2

THE FIRST DROP OF RAIN ON THE NASCENT EARTH

Earth is formed

The visitor enters the Natural History Museum, Vienna, up a wide staircase flanked by a painting of Kaiser Franz I, resplendent in red gown and breeches. The Kaiser is seated beside a fossil ammonite and he holds a green emerald that symbolizes his famous collection of 30,000 specimens. Behind him stands a smirking Johann von Baillou, who sold him the collection in 1750, while the Director of the Imperial Mint carries a tray dotted with specimens like buns on a baking dish. Near the staircase are the ammonite and emerald themselves, as well as the stuffed lapdog of the Emperor's wife.

The lavish museum demonstrates how seriously Austrian society regarded science. As ship-borne adventurers returned to Europe with specimens from distant continents, museums became exotic repositories of knowledge as libraries had been in the ancient world. The Vienna mineral halls display copper from the Americas, lapis lazuli from Afghanistan, and jade from the Far East.

But Hall 5 is extraordinary beyond measure. The room is crammed with misshapen meteorites, some as big as armchairs—the oldest objects on Earth. Surprised by the uncompromising brutality of the meteorites, teenage lovers recoil in the doorway and hug each other more tightly. Cold and remote, these brooding rock fragments are an extraordinary gift from the past.

Among the 8000 specimens are stony and iron meteorites from the cores of long-exploded planets, as well as meteorites with once-molten droplets of primordial matter. Visitors snap photos of each other beside a fragment of the meteorite that blasted out the Arizona Meteor Crater, their arms embracing its knobbly shoulder as though hugging an aged relative. A cabinet contains meteorites found in marine limestones by Swedish quarrymen: exploding on impact, the fragments plummeted to their tombs among the burrows of the ocean floor. A museum guide points

out a meteorite that fell in his French village in 1872 and was probably seen by his ancestors. And there is a meteorite that killed a dog in Egypt, its gas pockets matching the atmosphere of Mars, and one of only a hundred known fragments from the Red Planet.

Meteorites are mute witnesses to the first days of the Solar System.[1] A cloud of dust and gas on the edge of a spiral galaxy collapsed to form a proto-Sun surrounded by a disc of coalescing particles that are represented by the Vienna meteorites. Dating back more than 4.5 billion years, they also approximate the age of the Earth. A Mars-sized body later collided with the proto-Earth to generate a disc of molten rock and vapour that, over perhaps little more than a human lifetime, coalesced into the proto-Moon. Moon rocks yield ages nearly as old as the meteorites.

On Earth, radioactive heating produced oceans of molten magma, while meteorites, asteroids, and comets bombarded the surface. The early Earth was seemingly a hellish place, and the period before four billion years ago is named the Hadean Eon for Hades, the Greek god of the underworld.

And then the first drop of rain fell on the nascent Earth.

The first drop of rain

When did the first rain fall? In 1927, Professor Assar Hadding posed this surprising question to a meeting of the Royal Physiographic Society of Lund in Sweden. Hadding was a master in several branches of geology and was widely respected for his technical skill and imagination.[2] In the days before much radiometric dating, Hadding considered that the rocks of Sweden were among the oldest sedimentary remnants on Earth. Liquid water must have existed to lay down the original sediments, and this implied rain or dew and cycles of weathering and river transport. Although much older rocks are now known, Hadding's written account from the lecture vividly evokes the early Earth.

An ash-covered earth, no traces of water, no traces of life. A hot surface of earth under a hot atmosphere, rich in water-vapour and no doubt also in ash-dust. Volcanic cones and folding-mountains of greater height than the present ones, tablelands (future continental platforms) and wide depressions (future oceans), equally void and ash-coloured. Thus may the earth be pictured before the condensation of water. Then the first rains fall. What a wonderful impression is made on us by that which now takes place! Falling drops hit a ground never touched by water, a ground so hot that it has not been able to hold any liquid water. What salts were not to be leached out! How rapidly could not the rills cut deep grooves in the ash! Never have the rivers been so full of mud as during this first condensation.[3]

As Hadding reminded his audience, rain means rivers. But if rain had fallen on the early Earth, why not also on other planets?[4] In 1922 a meteorite fell into a frozen lake in Sweden, but Hadding failed to find it after the lake was drained. Three years later, a bright meteor crossed Sweden, and word came that a stone had fallen near the farm of Bleckenstad. Hadding's assistant returned with specimens and a dramatic account from a respected farmer who had seen a white ball of rock sweep over the heads of his niece and nephew and shatter on the ground. Scared out of his wits, the boy shouted that they should go home because the moon was falling. The farmer collected white limestone splinters that were scattered across the field, and Hadding found indefinable shell fragments in the limestone, which did not appear to match the local rocks.

After years of soul-searching, Hadding gave another lecture to the Royal Physiographic Society in 1939 entitled 'We and the World Outside'.[5] He reasoned that, although most meteorites come from the fragmented interiors of celestial bodies, such bodies might have contained a thin outer crust with Earth-like sandstone and limestone, laid down by water suitable for life. Might the Bleckenstad shell fragments represent extra-terrestrial life? Could the unremarkable boulders of the drained pond have included an Earth-like meteorite that he failed to recognize? We wait for new messages from the world outside, he concluded.

The Bleckenstad stone was not related to the meteor track and remains a mysterious 'pseudometeorite'. But Hadding was not afraid to present an interpretation that ran counter to perceived wisdom. And we now know that rain fell on Mars more than three billion years ago.[6]

Rivers more than four billion years ago in Australia

Earth's earliest rivers may be recorded in grains of sand. Under the Western Australian desert lie fluvial conglomerates and sandstones more than three billion years old, part of the Jack Hills belt in the Yilgarn Craton, the remnant of an ancient continent.[7] The rocks contain grains of the tough mineral zircon, eroded from much older rocks and transported by the rivers. Researchers have dated an extraordinary 100,000 of these zircon grains, 50,000 of which are more than four billion years old and one of which is 4.4 billion years old, the oldest known piece of the Earth (Fig. 2.1). Common in granitic rocks, these ancient zircons show that continents had formed surprisingly soon after the

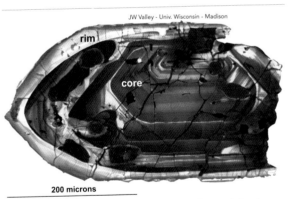

Figure 2.1 Zircon crystal from quartzite rock in the Jack Hills, Western Australia, shown in a colourized cathodoluminescence image. A date of 4.4 billion years based on conventional U-Pb geochronology, confirmed by additional analysis, shows that the crystal is the oldest known piece of the Earth. The date confirms that a crust existed prior to 4.4 billion years ago, which supports evidence for a cool early Earth that was habitable for life by 4.3 billion years ago. Information about the crystal from Valley et al. (2015). The image is used courtesy of John Valley at University of Wisconsin-Madison.

Earth accreted, and the composition of oxygen in the zircons suggests a cool surface and, probably, rain.

This was not a hellish Hadean Earth after all. Hadding's first drops of rain reach back almost to the beginning, carving channels through the loose volcanic debris as he envisaged. And if there was rain there were rivers and oceans, although we may never find their traces. A few whole rocks that date back more than four billion years may be remnants of these early river sediments, but after more than a century of age dating, they are the only direct evidence for the inscrutable Hadean Eon on a planet that repeatedly reworks its surface through volcanic activity and erosion. In contrast, six lunar landings over two weeks by the Apollo Mission returned Moon rock specimens more than 4.4 billion years old.[8]

Was there life—that most extraordinary of phenomena—in these earliest rivers and oceans?[9] The zircon evidence shows that the Earth's surface was suitable for life some 4.4 billion years ago. Chemical evidence suggests that living organisms were present four billion years ago, and probable microbial carbon is present in the sediments of soils and lakes nearly three billion years old. Comets and asteroids may have seeded the early Earth and its first rivers with simple organisms from elsewhere in the galaxy—the hypothesis of *panspermia*.

The first life may have inhabited hot springs, craters, and saline lagoons. Charles Darwin speculated whimsically about a 'warm little pond' where proteins formed to build a living organism.[10] After Hadding's writings were published in the United States, newspapers quoted his view that life originated in hot puddles in volcanic ash.[11] The Milwaukee Sentinel chortled 'life started in hot water' in 1931. But the Highland Recorder of Monterey, Virginia commented sagely on Christmas Day 1931 that what is important about life is not how it began but how we make use of it while we have it.

What might Earth's rivers have looked like four billion years ago? Clues are provided in modern Iceland where a thick lava crust has formed over the past fifty million years, with local granitic rocks.[12] Disregard an oxygen-rich atmosphere, moss and grass washing up lava cliffs like a green wave, and the brilliant air full of seabirds.

From the summit of a huge volcano, the faint young sun appears briefly through thick cloud, a dim version of what it will be in our day, and an occasional meteor burns out noisily overhead. The black basalt plain below is covered with wrinkled lavas that flowed smoothly from fissures and craggy flows that erupted explosively. A fierce wind whips up the black volcanic sand, and pools of boiling mud are dark with sulphides. On the horizon, the ocean waves have circled the globe and are breaking against the black landmass of the first continental crust.

Now that the atmosphere has cooled below the boiling point, hot-water rivers are running down volcanic cones that shine with gypsum crystals and clay as the rock begins to weather. Where rain has gullied the volcano, landslides have gouged out channels choked with ash and debris, and a persistent thread of water catches the sun where a braided river trickles between black gravel bars. The lava plain has little organized drainage: nothing is regular, nothing is linear. The river reaches the sea in a broad delta where the distinction between black land and dark sea is blurred.

The sun is going down in a brilliant red sunset through the volcanic dust and a very young moon is rising over the sea. As dusk descends, green patches glimmer among the lava flows and hot springs. Could they be microbes evolving in Hadding's ashy puddles or Darwin's warm little ponds? Sadly, they are too far away to be certain.

Cratons, supercontinents, and large rivers more than three billion years ago

Grains of sand are one thing, but large rivers require large continents. Across the Hadean Earth, plumes of magma domed up the surface and poured out vast lava fields, while rivers flowed across small, unstable regions of continental crust. Half a billion years later, the Earth has entered a smouldering adolescence.[13] The first stable *cratons* of continental crust (a term derived from the Greek *kratos*, strength) have appeared in what will become Africa, where floods laid down the world's oldest preserved river sediments of the Moodies Group, 3.2 billion years old in the Archean, on the Kaapvaal Craton.[14]

Archean rivers came dramatically to public attention in 1886 when prospector George Harrison found gold in the Witwatersrand range of South Africa.[15] A rough man with a troubled history, Harrison found a weathered outcrop or 'reef' of conglomerate and is said to have crushed a sample on a ploughshare in a widow's kitchen, concentrating the denser gold in her frying pan. Harrison was credited with discovering the Main Reef of the Rand goldfields, but some months later he sold his claim for £10 and wandered away, never to be heard of again.

Harrison may have assumed that he had found gold in young river gravels tilted by earth movements, for no one supposed that *ancient* river strata might contain gold eroded from bedrock. But such was the case. Known as 'banket' from its resemblance to a Dutch almond cake, the conglomerate dipped steeply below the upland surface and can be traced underground over an extraordinary 200 square kilometres. The Rand has yielded half of all the gold mined on Earth, owing its wealth to gold-rich river conglomerates and sandstones a few metres thick, laid down in broad, shallow river channels.

Soon the gold rush was in full swing and mines were operating all along the 'Randscape'. The wild mining capital of Johannesburg was described as a combination of Sodom, Gomorrah, and Monte Carlo, lacking anyone who could distinguish a violin from a vegetable.

The Rand conglomerates are extraordinary for other reasons.[16] They are 2.7 to 3 billion years old, laid down on the Kaapvaal Craton and among the most impressive braided-river strata on Earth. Fast-flowing water concentrated the dense gold nuggets along with larger pebbles of similar weight, and flakes of gold washed across sand flats and over the slipfaces of dunes and ripples. Dissolved gold was absorbed by living cells or precipitated as delicate films on microbial mats in the river pools.

A supremely significant threshold dawned some three billion years ago as the Archean drew to a close.[17] Continental crust was now extensive, and the underlying mantle was convecting slowly, with zones of weakness that allowed oceanic rocks to be drawn deeply under the continents. Driven by horizontal forces, the lithospheric plates with their continents began to drift inexorably across the Earth at rates of centimetres a year, colliding to form *supercontinents* crossed by mountain ranges. Plate Tectonics as we know it (see Chapter 4) had arrived, and the Earth achieved a tenuous maturity.

Nuna was among the oldest supercontinents, formed nearly two billion years ago around a mountain range from which rivers on the scale of the Amazon fanned out across North America.[18] A billion years ago, the supercontinent of Rodinia came into existence (Fig. 2.2), centred on the long-vanished Grenville Mountains that stretched from Labrador to Texas and rivalled the Himalaya and the Tibetan Plateau

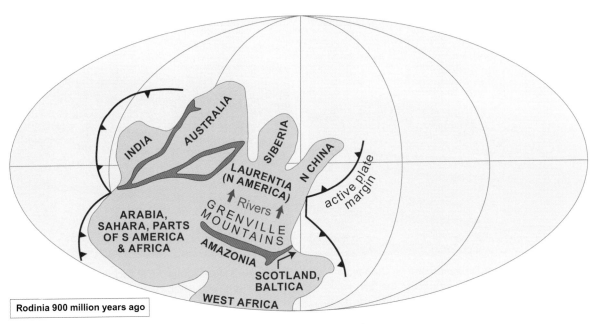

Figure 2.2 The supercontinent Rodinia, 900 million years ago, with large river tracts extending from the Grenville Mountains. After Li et al., 2008.

in scale.[19] With its enormous undivided landmass, Rodinia sourced larger rivers and more extensive spreads of sand and gravel than any river tracts on the modern Earth with its smaller, dispersed continents. Among the spectacular river deposits of Rodinia was the Torridonian Sandstone of the Scottish Highlands.

The Torridonian rivers of Scotland one billion years ago

The mountain of Quinag is built from Torridonian Sandstone, the sand of a river system that crossed Scotland during the Proterozoic, a billion years ago (Fig. 2.3A). Brown water oozes from the bogs through the sandstone outcrops, and heather gives way to rock along the formidable cliffs. The wind has freshened to gale force, sweeping the ravens over a dripping landscape that is unmistakably glacial with its scooped corries and hanging valleys. The only human presence is a lone figure near the summit.

During the Pleistocene Ice Age, glaciers cut through the Torridonian strata to the old resistant rocks of the Lewisian Gneiss below, some three billion years old. The remnant Torridonian strata rear up as steep-sided mountains from an Archean land surface that is almost as it was when the first Torridonian rivers swept the land. Where the red sandstone drapes the gneiss below the great bulk of Quinag, the visitor can touch an ancient unconformity—a gap in time of two billion years. Along the coast where broken fishing gear and the wreck of a seagull lie in the uncanny brilliance of the northern summer, the gneiss is veined where red sand and rock fragments tumbled into fissures in the deeply weathered surface. This is one of the older palaeo-surfaces on Earth, protected under a thick blanket of river sediment for almost a billion years.

In the cliffs of Quinag the sandstone beds reveal the river workings.[20] No deep channel forms are visible here, and the cross-bedded sandstones form thin sheets, some with twisted layers disrupted by ancient earthquakes. The rivers traversed a landscape dotted with wind-blown dunes and wrinkled microbial mats, where lake sediments preserved the delicate remains of organized cells with a nucleus, the earliest terrestrial eukaryotes.[21] The Torridonian channels were among a suite of river systems that fanned out across Rodinia from the Grenville Mountains, crossing the Canadian Shield to reach Scandinavia, South America, and Siberia.[22] In the imagination of geologists Rob Rainbird and Grant Young, the view from the Grenville peaks would have been extraordinary—an endless river plain unrelieved by a single plant or animal, with only the sighing of the wind and the tumult of the rivers for company.[23]

The rock ledges of the modern sea cliffs on Handa Island near Quinag are white with guano from thousands of nesting seabirds (Fig. 2.3B). The ledges follow the flat surfaces of the sandstone beds where pointed eggs can lodge without rolling into the breakers below. At the

Figure 2.3A Torridonian Sandstone on Quinag, northwest Scotland, with outcrops of fluvial sandstone about one billion years old. The Lewisian Gneiss, about three billion years old, forms the low ground at back left, with a profound unconformity between the two rock groups.

Figure 2.3B Torridonian sea-stack on Handa Island, with seabird nests on ledges formed by flat-lying river strata.

limit of sight is Cape Wrath, where the sandstones form the highest sea cliffs of the British mainland. A track runs to a remote lighthouse built by the grandfather of writer Robert Louis Stevenson, past old telegraph posts with cut, tangled wires: 'Wifi', says the minibus driver dryly, adding 'We are ahead of the times'.

Inland at Inchnadamph, the Torridonian strata are overlain by metamorphic rocks of the Moine Schist, seemingly much older. Charles Lapworth and survey geologists Benjamin Peach and John Horne investigated this apparent conundrum and showed that the Moine Schist was part of a *thrust sheet*, forced far to the west over the Torridonian strata along a gently dipping fault, the Moine Thrust. So intense was the debate that Lapworth suffered a mental breakdown and had to be invalided out, experiencing nightmares in which the Moine Thrust grated across his body as he lay in bed.[24] A photo of Peach and Horne at Inchnadamph shows the two men bent over their walking sticks like shepherds. So close was their working relationship that each knew the other's thoughts, and no one can tell what each contributed to their famous 1907 report. Later work on relatively unaltered regions of the Moine Schist suggests that the rocks were originally equivalents of the Torridonian Sandstone, laid down in the same basin. They were later buried deeply, metamorphosed under elevated temperature and pressure, and thrust westward over the unaltered Torridonian Sandstone at Inchnadamph.

The nearby hotel is a site of geological pilgrimage. Framed on the wall is the hotel register for a 1912 field excursion with the names of Peach and Horne heading the list. There is the large neat writing of Albert Heim from Zurich, whose Alpine work inspired Lapworth, and the signatures of Scottish researchers E.B. Bailey and C.B. Crampton—terrible blotters who found the unreliable hotel pen at its worst.

Extraterrestrial rivers three to four billion years ago

I spend my 58th birthday on Mars. More exactly, I spend the day in Becky Williams' basement in Wisconsin, which has a good view of Mars on her computer screen. Becky has studied Martian landscapes for years and would go to Mars if she could. Her daughter knows the 'favourite planet' almost as well as the neighbouring streets.

In the seventeenth century, astronomers trained their telescopes on Mars and realized that Mars was a world like ours, and that a plurality of worlds lay in the stars beyond, some perhaps inhabited.[25] In the late nineteenth century Giovanni Schiaparelli identified dark streaks as water bodies or 'canali' (channels), and the American astronomer Percival Lowell inferred that the 'canals' were bordered by irrigated zones of vegetation planted by intelligent life. Writers picked up the imagery: the frozen canals where schoolboys skated to safety with the little Martian creature Willis in Robert Heinlein's *Red Planet*; the glittering canals with drifting boats in Ray Bradbury's *The Martian Chronicles*, a good place to drop empty beer bottles.

When Mariner 9 moved into Mars orbit in 1971, the 'canals' were found to be dark streaks of windblown dust, leading Carl Sagan to wonder which end of Lowell's telescope had shown intelligence. But the surface of Mars also displays river channels, dry and frozen in time, that flowed more than three billion years ago during the Hadean and Archean eons on Earth, before the Moodies and Rand rivers traversed Africa.[26] Rocks blasted to Earth from these Martian landscapes may be Hadding's messengers from the world outside, but as far as we know, his early rain fell on a lifeless landscape.

Figure 2.4 Canyon of Valles Marineris on Mars. The canyon is more than 4000 km long (about the width of the United States), 600 km wide, and up to 8 km deep, and it was cut about 3.5 billion years ago. Courtesy of NASA/JPL-Caltech/MSSS.

Figure 2.5 River and delta strata 3.6 to 3.8 billion years old in Gale Crater on Mars, dipping towards lake strata in the distance. Note cross-bedding in a rock at lower right. Photographed in 2015 by Curiosity Rover, reproduced courtesy of NASA/JPL-Caltech/MSSS.

Dwarfing Earth's Grand Canyon is Valles Marineris, 4000 km long, 600 km wide, and up to eight kilometres deep, dating back at least 3.5 billion years and the largest canyon in the Solar System (Fig. 2.4). In the Eberswalde Crater some 3.5 billion years ago, fan-shaped river deltas built into a lake, with distributary channels and meandering river channels with cutoffs and scroll bars, the channel banks strengthened by stiff clay, precipitated minerals, or ice. The lovable little Curiosity Rover has sent back images of delta and lake strata 3.2 to 3.6 billion years old from Gale Crater (Fig. 2.5), as well as photos of conglomerates with rounded pebbles, cross-beds, rippled layers, and unconformities. And in the Arabia Terra area, a ridge nearly 100 km long and 1 km wide is interpreted as an 'inverted' channel fill, left standing after the softer surrounding sediments were eroded. Estimated to be 3.9 billion years old, the ridge may be the oldest preserved river sediment in the Solar System.

Most remarkable of all are the outflow channels, far larger than those associated with the Ice Age on Earth (see Chapter 13). With dry waterfalls that dwarf Niagara, they were cut by cataclysmic floods about 3.7 to 2.6 billion years ago, when groundwater erupted through collapsing frozen ground, brilliantly depicted by Kim Stanley Robinson in *Red Mars*. The flow rates may have exceeded 100 million cubic metres a second—nearly 300 times the flow rate of the largest Amazon flood ever recorded.

On Titan, the largest moon of Saturn, cloudbursts generate floods of liquid methane on an icy crust far below the freezing point of water. The Huygens probe that landed on the surface in 2005 descended through a hydrocarbon haze under a parachute made by the textile factory in my hometown of Tiverton—the furthest landed human object to date. Braided and perhaps anastomosing rivers are there, along with lakes and dunes of hydrocarbons, a petroleum engineer's dream.[27]

A Universe of rivers

The Universe is unfolding in Don Wightman's art studio near Toronto, a converted garage where the walls are adorned with photos of a spiral galaxy and cross-beds taken by the Mars Opportunity Rover. Near a trolley with a scatter of brushes and paint tubes stands an easel bearing a large canvas—*Sun Nursery*, a pulsating acrylic riot of colours and shapes emerging from blackness (Fig. 2.6). There are white, blue (too much of this, says Don), red and yellow stars, and swirls like string theory. Standing on a ladder to reach the highest stars, Don works with a small frame until he is satisfied with each area. Don—an aircraft engineering apprentice who, while waiting for a bus to math class, fell into conversation with art students and, on a whim, took a bus with them to the art college and eventually to the far reaches of the Universe.

Figure 2.6 Sun Nursery by Don Wightman, acrylic on canvas, unfinished. Courtesy of Don Wightman and Margie Wightman.

There will be planets in the *Sun Nursery*. Some will be barren but on others the first chemical rain will be falling, driving river systems and perhaps encouraging life. Thousands of exoplanets have been discovered and our galaxy may have 100 billion more.[28] Some five billion years from now, our sun will swell into a red giant and terminate the river histories of Earth, Mars, and Titan. But a wealth of other River Planets will be there to take their place.

CHAPTER 3

HOW PLANTS BENT AND SPLIT RIVERS

A vegetated world

As a child, I found plants to be tediously immobile, unlike the walking killer plants of John Wyndham's novel *The Day of the Triffids*. My father, however, often sat and watched plants, whether they moved or not. But there were exceptions to my disinterest. Wistman's Wood on Dartmoor was a magical place, a remnant of the native oak woods that once clothed Britain. The low gnarled trees, some as much as 400 years old, were twisted and bent, their branches covered with epiphytic ferns and lichen. And there was the Ankerwycke Yew at Runnymede near London, eight metres in circumference, which witnessed the signing of Magna Carta and may once have housed a hermit in its hollow trunk. King Henry the Eighth courted Anne Boleyn under its boughs before eliminating both Anne and the nearby priory.

Few places display the raw ability of simple vegetation to colonize rock better than Iceland. Little more than 200 years after the catastrophic Laki eruption of 1783, pale green moss shrouds the rubbly lava as far as the eye can see. But I learned most about the link between vegetation and rivers from two North American forests. Along the Saskatchewan River of northern Canada (see Chapter 19), deep roots protect the undercut banks and hold the large poplar trees in place, their trunks tilted towards the channels, while willows grow so profusely along the islands that a boat cannot reach the bank. And Muir Woods near San Francisco is a remnant of the once-vast redwood forests that John Muir in his essay *The American Forests* considered the finest ever planted by God (Fig. 3.1A).[1] Visitors fall silent among the shafts of light, awed by these living symbols of endurance that were old in Julius Caesar's day. The standing giants—too tall to be seen fully through the dense canopy—anchor the riverbanks, where a tiny wren, among the smallest of birds, hops and chatters on the toppled trunks that jam the winding creeks.

This may be the closest that I shall come to imagining the forests of the Devonian and Carboniferous, when vegetation changed rivers for ever. As in Far North Queensland, where groves of trees cover the river channels and sandbars (Fig. 3.1B), forests and rivers may have been barely distinguishable during the dry season. But farmers from the Neolithic onwards cut down and burned the riverbank and floodplain trees to make way for pasture and cropland, and humans are now the geological river agents *par excellence*.[2] Fallen trees and logs filled North America's rivers until the Civil War, shaping channels and providing fish habitat, until millions of drifted trees and logs were removed from the Mississippi and other rivers to improve navigation.[3] Blasted out in the 1830s, the 'Great Raft' on the Red and Atchafalaya rivers of Louisiana was a log jam 250 km long that had taken hundreds of years to form. And as part of a lucrative timber industry that has largely passed out of knowledge, mounds of cut logs were released with dynamite into river channels during spring floods, jamming up on rocks and waterfalls en route to water-powered sawmills and choking the channels with logs, bark, and sawdust. With the riparian trees went the stabilizing roots that resisted bank collapse, and soil erosion washed away the fertile community of rooted plants, microbes, fungi, and animals that had dominated river floodplains ever since the Great Green Landscape Revolution more than 400 million years ago.

The Great Green Landscape Revolution of the Paleozoic

River plains and rocky surfaces carried films of microbes during the Archean and Proterozoic, but these

Figure 3.1A Redwood trees at Muir Woods National Monument, California.

Figure 3.1B Trees growing in dry channels and on sand bars at low-stage in the Gilbert River, Australia.

organisms appear to have had little effect on the rivers. Then during the Ordovician some 470 million years ago, plants related to modern mosses and liverworts evolved short upright stems and supporting vascular tissue for the transport of water and nutrients.[4] By 410 million years ago in the Early Devonian, wetland biomass was sufficient to form the first peats, turned to coal through geological time—the energy of the sun trapped in fossilized vegetation. Wildfires charred the plants and left the first charcoal. By 385 million years ago in the Middle Devonian, forests contained trees more than 20 m tall with radiating roots and an understorey of smaller plants and creepers.

During the Carboniferous Coal Age that followed, early conifers, club mosses, horsetails, and ferns left a wealth of fossils in river and wetland sediments. For the first time in Earth history, peat covered large parts of the continents and drought-resistant conifers with deep roots colonized plains and uplands. Newly evolved arthropods nibbled on leaves and vegetable mould, and early amphibians and reptiles left their footprints in dry channels. The barren landscapes of the Proterozoic gave way to a rich ecosystem, and the Earth passed a point of no return.

The effects of the Great Green Landscape Revolution were enormous and global. Whereas the Proterozoic floodplains developed short-lived soils with little clay, the simple plants and rooted vegetation of the Silurian and Devonian fragmented the rocks, while organic acids broke down feldspars and other unstable minerals to clays with the same chemical components. As weathering intensified, mountains were eroded at an unprecedented rate, rivers transported sediment and nutrients to the ocean, and plankton showered the seabed, burying vast amounts of carbon. So dramatic was the loss of atmospheric carbon dioxide—our familiar greenhouse gas—through weathering of rocks and burial in wetlands, lakes, and oceans that, by the end of the Devonian, the Earth was plunged into an ice age that continued through the Carboniferous and into the Permian. As photosynthetic plants covered the land, atmospheric oxygen levels rose.

And in little more than 100 million years, rivers changed out of all recognition. Plants and mud strengthened the riverbanks so effectively that many channels become sinuous and 'bent' (meandering), while others divided around vegetated islands and 'split' (anastomosing). Where wide sandy channels had once crossed the plains, a mix of braided, meandering, and anastomosing rivers now wound through forests and wetlands. Animals evolved within ecospace that had not existed before—floodplain soils with leaf litter, forested riverbanks, and the log jams that favoured aquatic life. By the end of the Carboniferous, the Earth's river landscapes would have been much as the Neolithic farmers found them. And some of the first studies of Devonian and Carboniferous vegetated river landscapes were made by a remarkable group of enthusiasts in Scotland and Canada within the decades following James Hutton's discovery of geological time.[5]

Scottish geologists find rivers in the Old Red Sandstone

Hugh Miller was born at Cromarty in Scotland in 1802, oddly misshapen and destined, thought the midwife, to be the village idiot. His grandfather had been a pirate and his uncle had collected fossils under fire during the Battle of the Nile. A wild red-haired teenager, Miller attended a tough school that boasted an annual cock fight, and he is recorded as knifing a fellow student and fighting with the teacher. Apprenticed as a stonemason at age seventeen, Miller was fascinated during his first day's work to find the scales of fossil fish in a rock slab, along with wood fragments and ripple marks.[6] For a time, he carved masonry for a stately home in Edinburgh but, afflicted with silicosis from inhaling stone dust, Miller returned to Cromarty where he inscribed gravestones and discovered a vocation in writing. He spent his free time collecting fossil fish from the cliffs of the Old Red Sandstone (Fig. 3.2), at that time a largely unstudied Devonian rock formation.

Miller moved to Edinburgh as the editor of a religious magazine for which he wrote articles about Devonian fossils. He published *The Old Red Sandstone* in 1841, describing in imaginative prose the ancient landscapes and their creatures, and the book went through 26 editions, praised by admirers who included Charles Dickens. Miller explored the link between Earth history and religious views, accepting the vast duration of geological time but refuting an accidental origin for creatures. His later book *The Footprints of the Creator* was a sensation, and in 1854 he lectured to a crowd of thousands in London. But overwork, silicosis, and possibly a brain tumour damaged his health, and he took his own life in 1856. Mourners crowded the streets as his funeral cortege passed.

In 1845 Miller received a letter from Robert Dick, a baker at Thurso in the Caithness region of northern Scotland, along with samples of a gigantic Devonian fish. Dick was fascinated by geology and botany, and he painted Egyptian deities on the walls of the bakehouse where he entertained guests. In a later letter to Miller he noted that the local building stone was so fossiliferous that Thurso was virtually made of dead fish. A privately religious man but a free thinker, Dick had no hesitation in telling Miller when he was twisting facts to fit religious ideas. 'How can I or any man,' he commented 'while looking at a plant, say that it has maintained all

its particular characteristics *unchanged* since it came from the hands of its Maker?' He thought, however, that 'first stocks must have had a Creator. They could not spring up out of the ground *unbidden*.'[7]

The eccentric baker with his chimney-pot hat and swallow-tailed coat often walked through the night and collected fossils by day, returning for the evening's baking. A lover of flowers, he planted unusual varieties as he walked to ensure their survival. He was mistaken for a salmon poacher and was nearly shot by an inexpert rifleman. Was he digging for silver or burying gold, the locals wondered? A self-effacing man who did not promote his work as Miller did, he was nevertheless a prolific correspondent. How long had it taken for the Devonian strata to form? 'These are pertinent questions, – questions which enter one's very soul. Then man feels instinctively his own littleness, and his utter inadequacy to solve even the simplest of his questionings.'[8]

Distinguished visitors came to the bakehouse. Roderick Murchison, President of the Royal Geographical Society, recalled Dick explaining the geological features of Caithness by moulding dough on a baking board. When Murchison sought to buy some of his fossils, Dick insisted on presenting them as gifts, despite failing health and business losses (he was forced to sell his precious fossils after losing an assignment of flour in a shipwreck). The plants of the Old Red Sandstone were later made famous by coastguard Charles Peach, a friend of Dick and father of Benjamin Peach of Moine Thrust fame. Peach found a beautiful fossil plant with leaves and the coaly remains of bark, after which 'his mind ran after plants' across the Scottish outcrops.

In 1912 physician William Mackie found previously unknown plants in blocks of crystalline silica or chert in a stone wall at Rhynie near Aberdeen. Trenches showed that the blocks came from the Old Red Sandstone, which lay below the fields. Hot springs had entombed the growing plants in silica crystals, and floods of river sediment had buried the petrified Devonian vegetation. The site might have been lost without the intervention of Geoffrey Lyon, a reclusive botany professor who retired to Rhynie and purchased the field, bequeathing it for agriculture and scientific research along with substantial funds. When visitors sought fossil souvenirs, Lyon recorded uncomplimentary notes about them.

The Rhynie chert, 400 million years old, is an

extraordinary time capsule of a primitive ecosystem. On cut and polished surfaces of the chert are black lines like pen strokes, the preserved stems of *Rhynia*, along with dark layers of plant litter and microbial films (Fig. 3.3). Researchers have found fungal parasites and lichens in the chert, along with damaged plants that repaired their wounds. There are harvestman spiders with their spindly legs, centipedes, and minute spider relatives with air-breathing, book-like lungs. Remarkably preserved on the Devonian river plains that Miller, Dick and Peach made famous, Rhynie lies at the root of all forests and land animals.

Devonian and Carboniferous rivers and vegetation in Canada

William Logan was a manager in his uncle's smelting works in Wales. Growing interested in the Carboniferous coal that fuelled the furnaces, he discovered upright trees preserved with the soils in which they grew. In 1842 Logan became the first Director of the Geological Survey of Canada with a mandate to explore the resources of the nation, then comprising southern Ontario and Québec. The British crown had suffered the crippling loss of the American coalfields after the War of Independence and needed coal resources.

Logan surveyed the Gaspé Peninsula of eastern Québec where he worked with First Nations inhabitant John Basque, staying at times in Basque's wigwam where they dined on porcupine and bear. He worked with his head in a gauze bag to keep off the blackflies and, raggedly dressed and peering through surveying equipment, he was considered a lunatic by local fishermen. Logan found petroleum oozing from the beaches, and he discovered bitumen in the rocks at Tar Point where adventurer Jacques Cartier had tarred his ships three centuries earlier. Logan deduced that the oil had gathered under the arch of a geological fold. But although he discovered splendid plant fossils in the ancient river strata, he found only a few thin coal seams.

Logan showed his 'fossil vegetables' to John William Dawson, then in his early twenties, who had first collected fossils while excavating slates to make pencils behind his Nova Scotia schoolhouse. He correctly identified Logan's fossils as Devonian, a time of modest biomass and the earliest thin coals in comparison with the thick and abundant Carboniferous coals of the

Figure 3.2 Fluvial strata of the Old Red Sandstone, of Devonian age, in the Old Man of Hoy, a seastack 137 m high in Orkney, Scotland.

Figure 3.3 Early Devonian Rhynie Chert, with simple stems of *Rhynia*, an early vascular plant, enclosed in pale hotspring silica. Coin 3 mm thick for scale.

lost American colonies. He also noted a fragment of a curious small tree.[9] Dawson later visited Gaspé and was impressed by the enormous cliffs of river strata where, perennially observant, he noted rills made by rolling rain drops from ancient storms. Finding more tree fragments, he named them *Prototaxites logani* and thought that they resembled a fungal mycelium. Dawson's 'small tree' was in fact a gigantic fungus nearly nine metres tall that had lived on plant litter and towered over the Devonian landscapes for fifty million years. Like the mythical triffids, *Prototaxites* was among the most bizarre organisms ever to inhabit the Earth.

Unusually in an age of museum experts, Dawson preferred to work with rock outcrops, observing Carboniferous river channels filled with drifted trees and interpreting the ecology of ancient landscapes. Later the Principal of McGill University, he was known for carrying soup and bread to sick students.

Perhaps better than anywhere else, the cliffs of Gaspé and other parts of eastern Canada display a landscape in transition as plants took root.[10] The braided-river sandstones are littered with plant fragments like coffee grounds, and they contain much more mud than their Archean and Proterozoic counterparts, a testimony to the impact of plants on weathering and soil formation. Trunks of *Prototaxites* in river sandstones are the oldest logs in the fossil record, and a rooted frond swung around by water or wind inscribed the earliest known scratch circle on the sand. Newly evolved millipedes pushed through decaying stems or burrowed in the soft soil. By the later Devonian, more than half of the world's river formations would be laid down by meandering rivers with banks strengthened by roots and mud.

Carboniferous rivers and forests are spectacularly seen in the Joggins Fossil Cliffs of eastern Canada, which form a wall of rock along the Bay of Fundy (Fig. 3.4A), home to the world's highest tides (16 m from low to high tide). Far out on the weed-covered sandstone 'reef' of Coalmine Point, the subtle turn of the tide can be heard on a still day, but within a few hours, the tide is hammering at the cliffs and tumbling rock fragments to the beach, where marvels hidden for three hundred million years are revealed (Fig. 3.4B).

In 1842, Charles Lyell observed 17 upright trees at Joggins (Fig. 3.5), many growing out of coal seams that represented the original peaty soil, and he was stunned to see a standing tree more than seven metres tall. He might never have seen anything more magnificent, Lyell thought.[11] Similarly captivated by the coal seams and trees, William Logan paced along 15 km of cliff in five days in the following year, working around the tides and taking a Sunday rest. He recorded in pencil in his notebook the number of paces (two feet nine inches) that represented each dipping bed at beach level, measuring their dip with a clinometer and—an accomplished artist—sketching the fossil trees. Back home he inked over the pencil record and used the paces and dips to record nearly 4.5 km of strata, reluctantly publishing the results in a government report ('who the devil ever reads a report').[12] His notebook also records the cost of silk cravats, *de rigueur* for Victorian geologists.

Figure 3.4A Joggins Fossil Cliffs, Nova Scotia, Canada, a thick succession of river and shallow marine strata of the Carboniferous Joggins Formation.

Figure 3.4B Wood in fallen blocks of channel sandstone, Joggins Formation, largely from lycopsid trees. Scale is 50 cm long.

Figure 3.5 Upright lycopsid trees in the Joggins Fossil Cliffs. The trees were entombed in their growing position when sand from shallow distributary channels swept through the wetland forests.

Most remarkable of all were the fossils that Lyell and Dawson excavated from the sandstone fill of a tree trunk. They found charcoal, amphibian bones, and the tiny snail *Dendropupa vetusta*, still the earliest known land snail. In another tree Dawson found the bones of the world's earliest reptile, the tiny *Hylonomus lyelli* or 'Lyell's forest dweller'. He used gunpowder to blow up part of Coalmine Point, removing 25 standing trees and finding the bones of some hundred creatures. Like the feral pigs of Australia, these early vertebrates probably lived in trees hollowed out by wildfires, an explanation suggested by Lyell and Dawson. A wanderer through the Joggins forests at dusk would have seen the glowing eyes of countless creatures peering from the trunks.

In his book *On the Origin of Species*, Darwin drew on Joggins in discussing the fossil record, noting 'we have the plainest evidence in great fossilized trees, still standing upright as they grew, of many long intervals of time and changes of level during the process of deposition.'[13] In his review of Darwin's book, Bishop Sam Wilberforce ('Soapy Sam') used *Dendropupa* to contest (unconvincingly) Darwin's evolutionary views:

The rare land shell found by Sir C. Lyell and Dr Dawson in North America affords a conclusive proof that in the carboniferous [sic] period such animals were most rare, and only the earliest of that sort created … if terrestrial animals abounded, why do we not see more of their remains than this miserable little dendropupa about a quarter of an inch long?[14]

Joggins probably featured in the 1860 evolution debate at the newly opened Oxford University Museum of Natural History. It was an ominous venue for the debate. When funds for stone carving ran out during the construction, the disgruntled stonemasons were dismissed after carving birds in a parody of Convocation and, as rumour had it, monkeys with the heads of university dignitaries. After acrimonious debate in which Soapy Sam probably waved a specimen of *Dendropupa*, Lady Brewster fainted, angry shouting broke out, and the protagonists went off cheerfully for

Figure 3.6 Fills of narrow fixed channels (arrowed, probably anastomosing rivers), 1 to 5 m thick, cut into red floodplain mudstone in the Carboniferous Tynemouth Creek Formation, New Brunswick, Canada.

dinner together. Scientists and clergy alike were divided on the evolution issue, but sites like Joggins could no longer be disregarded.

Joggins and strata of similar age yield a story of rivers evolving along with vegetation and animals.[15] Gone are the barren plains swept by sandy rivers. In their place are meandering channels and narrow anastomosing channels (Fig. 3.6)—among the world's oldest examples—that were locked in place by tough mud and the deep roots of newly evolved dryland trees, including early conifers. Some channels are choked with logs, a far cry from the sparse *Prototaxites* logs of the Devonian, and others contain sediment bars that built up around thickets of horsetails, the capable pioneering plants of the Carboniferous. Upright trees stand tilted on former riverbanks, and many vegetation features seen along modern rivers can be matched in the Joggins cliffs (Fig. 3.7).

The river channels were alive with organisms. Some contain the shells of large clams and the bones of early tetrapods: they may have been dry-season waterholes

like those of the Australian Channel Country. Wrinkled microbial mats are crossed by the trackways of *Arthropleura*, an enormous arthropod that scuttled into the channels to feast on decaying vegetation. The tetrapods left trackways on the channel floors, with prints that are minute or half a hand in size, and horseshoe crabs left the delicate impressions of their tails and appendages. And a dragonfly-like insect was preserved on a muddy floodplain surface dotted with rain prints.

As yet there were no colourful flowering plants, nor grasses with their exceptional ability to bind the floodplain soils: they did not appear until the Cretaceous. And later extinctions drastically reduced plant and animal communities, perhaps briefly setting rivers back almost to a pre-vegetational state. But Gaspé and Joggins mark the start.

Rivers come alive in the Devonian

To river specialist Luna Leopold, a river can be likened to a species, with each reach an individual in a

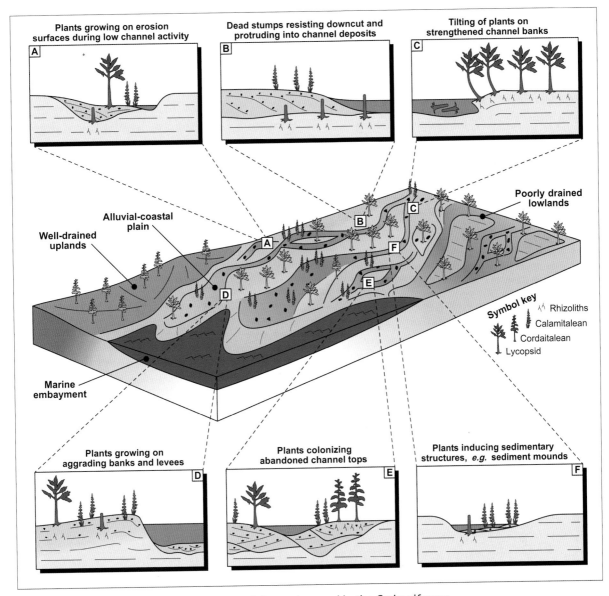

Figure 3.7 Relationships between plants and rivers observed in the Carboniferous Joggins Formation of Nova Scotia, Canada. After Ielpi et al., 2015.

population.[16] But Leopold stopped short of considering rivers to be alive. During the 1960s, James Lovelock formulated the Gaia Theory, named for the Greek goddess who personified the Earth.[17] He suggested that the Earth's atmosphere, rocks, oceans, and organisms had evolved together through the long tenancy of life, after early organisms charged the atmosphere with oxygen and regulated the Earth system to their advantage. In Lovelock's view, the Earth became an enormous organism where animate and inanimate matter were tightly coupled, like a living tree with a trunk of dead wood. The first photos of Earth from Space finally showed us a planet where life and matter are intimate partners. Encouraging his sceptical fellow scientists that they stood to lose only their research grants, he retired to rural Devon to think.

From a Gaian understanding, rivers can be considered living organisms that entered fully into the life of Gaia during the Devonian some 400 million years ago. Let us name the Earth's living rivers for *Anuket*, the Egyptian goddess and personification of the Nile with her headdress of river reeds or ostrich feathers, whose festival coincided with the yearly, nourishing flood of water and silt. Under her guidance the Earth has grown green with age.

Permian dryland rivers of Texas on the supercontinent Pangea

In the town of Seymour on the Texas ranchlands, 'Pray for Rain' signs stand outside banks and churches. Hardship is no stranger to a people habituated to droughts and scorching winds, where oil pump-jacks help keep the town alive and shotguns are 'snake-tamers' in rattlesnake country. The townsfolk are unfailingly courteous, neither inquisitive nor withdrawn. Restaurants are packed with folk in denim, some with cowboy hats and spurs, and a friendly rancher brings over a box of fossils. Back at the motel, a hunter waves a cheery greeting as he cleans a turkey on the tailgate of his pickup truck as feathers blow across the forecourt. I bog the four-wheel drive on the ranch road, leaving big mesquite spines sticking out of a tyre, which hisses flat as a mechanic pulls them out.

Through the nineteenth century, fossil hunters excavated bones by the thousand from the Seymour badlands, searching for the remains of creatures that President Thomas Jefferson (familiar with bone sites) predicted would be found living out West. In the Permian strata they found bones of the 'Texas Finback' *Dimetrodon*, a mammal-like reptile with a sail of spines, and little *Seymouria* with its broad skull and short legs. There were early sharks, the earliest-known frog, and burrows where lungfish and amphibians waited in vain for the rains that never came.

Descendants of the Joggins swamp creatures, these adaptable animals mastered the arid river plains on the western edge of Pangea, a supercontinent that included almost all land on Earth.

The red Permian strata contain meandering point bars (Fig. 3.8), and walking the plains is to traverse an ancestral river landscape where trickling water evaporated among mud cracks and roots.[18] Where the riverbank trees once grew, exquisite fossil fronds, their edges clipped by feeding insects, fell into the mud of abandoned channels (Fig. 3.9). With hammers in holsters, we clank across the plains like early reptiles, crawling up the badland gullies to peer at *Dimetrodon* bones and measure the flow directions of ancient ripples.

4 m

inclined point-bar deposits of silt/sandstone

conglomerate at channel base

inclined point-bar deposits

conglomerate at channel base

floodplain mudstone

channel deposits

Figure 3.8 Permian meandering-river strata with inclined point-bar surfaces in the Clear Fork Formation near Seymour, Texas.

Figure 3.9 Permian fossil (*Culmitzchia* sp.) from the fill of an abandoned channel in the Permian Clear Fork Formation of Texas. Courtesy of Smithsonian Institution, specimen PAL544670. Photo by M. Gibling.

Figure 3.10 Petrified forest in Arizona, with Triassic logs eroded from channel and floodplain strata.

The modern badlands form a parallel to that long-gone world. Winding creeks, cracked by drought, traverse a wilderness of dryland vegetation where cowboys marshal the cattle. Cactus is everywhere, pink blossoms crawling with pollen-smothered insects. Animals descended from the dryland creatures of Pangea are our companions—roadrunners with necks and tails outstretched, javelinas, and the capable feral pigs that draw hog hunters to the area. The distant howls of coyotes are borne on the wind, and a prairie falcon wheels out of the west. In the afternoon the wind whips up a dust haze, a miniature version of the 1930s Dust Bowl that took Seymour's topsoil, and a gust of wind sends my notebook tumbling into a cactus bush. A rattlesnake is coiled up nearby, its erect tail rattling. 'Walk, walk quickly!' says my companion. I walk quickly.

A cow has collapsed, and the ranch manager is feeding her by hand as she lies slumped by his truck. He has lost 35 cows this year from lack of rain and grass. Small clouds drift by but no rain falls. Next day the cow is lying there lifeless.

Pangea came into being some 300 million years ago around a mountain range as big as the Andes, which stretched across Europe and down the Appalachians to Mexico.[19] For more than 100 million years, large rivers fed by monsoon rains and glaciers crossed ancestral North America through an arid landscape that, based on analysis of salt crystals, was hotter than any on Earth today.[20] Trees flourished along the rivers, preserved in the splendid Petrified Forest National Park of Arizona where Triassic trunks litter the landscape, eroded from the river strata (Fig. 3.10).

And then the supercontinent rifted apart and our modern rivers came into being—rivers with names.

PART 2: OUR MODERN RIVERS

CHAPTER 4

BREAKING PANGEA: THE ANCESTRAL RIVERS OF AFRICA

Alfred Wegener discovers Pangea

In 1910, a friend of German scientist Alfred Wegener showed him a new world atlas that he had received for Christmas. For hours they pored over maps that showed the apparent fit of Africa and South America. In a letter to Else Köppen, who would later become his wife, Wegener wrote 'I'm going to have to pursue this'.[1]

There was nothing new about the continental fit. Maps became increasingly accurate through the 16th century, and cosmographer Gerardus Mercator made world maps that showed the curiously matched continents, quizzing sailors in the port of Antwerp and drawing on the writings of Marco Polo and the *Geography* of Ptolemy of Alexandria from the second century CE.[2] The Indian Ocean had long been charted by Arab traders, and Chinese Admiral Zheng He sailed down the African coast in the fifteenth century, using maps from China and Korea and returning with a giraffe. Rivers and coastlines were soon depicted in unprecedented detail, and Malvolio in Shakespeare's *Twelfth Night* was observed to 'smile his face into more lines than is in the new map with the augmentation of the Indies'. After Mercator's death in 1594, more than a hundred of his maps were combined in an *atlas*, a name that he coined for the Titan who shouldered the celestial spheres. Two years later Abraham Ortelius, a colleague of Mercator, suggested that earthquakes and floods had split the Americas from Africa and Europe.[3]

But it was not the fit of the Atlantic coasts, which might only be an artefact of sea level, that drew Wegener's eye. The new atlas showed the water depth along the continental margins using information gathered by the 1870s Challenger Expedition. Africa and South America fitted together far *below* the water surface, and this implied something fundamental about continental structure.

Wegener was a meteorologist who had studied classics and science in Berlin and spent his summers climbing in the Alps and fencing in Heidelberg.[4] At the Royal Prussian Aeronautical Observatory, he raised kites and balloons to more than 14 km on cables, once narrowly escaping electrocution when lightning incinerated the wire. He made a record balloon flight of 52 hours, and he was knighted by the King of Denmark after crossing the Greenland Icecap, where he shot a polar bear that was coming for him. Above all, he knew muddled thinking when he encountered it.

In 1912, Wegener wrote about the continental fit. Most scientists considered that continents and oceans were permanently fixed but that the continents were at times connected by land bridges across which animals migrated. As far back as 1769, the clergyman-naturalist Gilbert White discussed the view that an isthmus had once crossed the Atlantic but had later broken down. He commented 'But this is making use of a violent piece of machinery; it is a difficulty worthy of the interposition of a god!', adding a quote from the Roman poet Horace '*Incredulus odi*' (being sceptical, I detest it).[5] A rigorous physical scientist, Wegener knew that the continental crust was less dense than the rock of the ocean floor and liable to take a higher elevation following the principle of *isostasy*. Continents and land bridges could not just sink out of sight below the waves. The fit implied that the continents had moved apart, buoyed up on slowly flowing denser rock below. If so, continental collision would have forced up mountain belts of compressed rock—and Albert Heim, a participant in the Moine Thrust excursion, had illustrated such compression in the Alps. Elsewhere, the Earth's crust was rifting apart, for example in the Rhine Valley near Heidelberg. Was the Atlantic Ocean an enormous, widened rift?

Wegener also drew on the Permian plant *Glossopteris* (Fig. 4.1), the curious distribution of which could be explained if Australia, India, South America, and Africa

[Left] **Figure 4.1** *Glossopteris* leaves from the Reid's Mistake Formation, Permian of Australia. The plant fossil is found in parts of Gondwanaland that are now widely dispersed.

[Below] **Figure 4.2** The supercontinent Pangea, 200 million years ago, with dates of breakup into separate continents. After Seton et al., 2012.

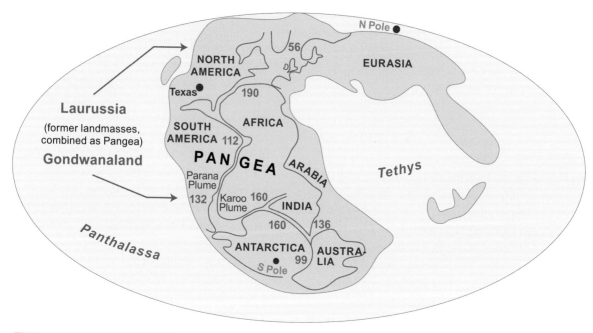

Pangea 200 million years ago

Dates show opening of oceans between continents in millions of years ago

had once been connected as part of Gondwanaland. Palaeobotanist Marie Stopes had urged Robert Falcon Scott to bring back plant fossils from Antarctica after he refused to take her with him, and *Glossopteris* fossils were in the rock samples that the ill-fated Scott and his companions dragged on sledges from the South Pole in 1913.[6] A member of the party that found Scott's body defended this fatal effort, commenting that 'these same specimens dated a continent and may elucidate the whole history of plant life.'[7] Antarctica, far distant from the other continents, was also part of Gondwanaland.

Wegener was called up for military service early in the First World War, but a spent bullet lodged in his neck as he led an infantry charge and he was later invalided out with a heart problem. While convalescing in 1915, he published *Origins of Continents and Oceans*, later running weather stations behind the front and teaching navigation to zeppelin captains. Working with Else's father, climatologist Wladimir Köppen, he published maps that tracked the continents back through time, and he named the reassembled supercontinent of the Carboniferous and Permian *Pangäa* ('All-Earth', later transliterated to *Pangea*), which included the components of Gondwanaland (Fig. 4.2). His continental reconstructions, among the most iconic images in geology, show rivers in their modern courses: it was for others to work out where and when they had once flowed. Wegener died of a heart attack on the Greenland Icecap in 1930.

Continental drift generated intense controversy among geologists who lacked Wegener's rigorous background. As late as 1944, American geologist Bailey Willis drew on volcanic land bridges to account for the scattered distribution of Gondwanaland fossils. Continental drift was *ein Märchen*, he declared, a fairy tale.[8] But post-war studies of rock magnetism confirmed that the Indian subcontinent had drifted north across the equator, and marine drilling showed that rocks of the modern ocean floor were less than 200 million years old, in contrast to ages of billions on the continents. The oceans had formed as Pangea drifted apart.

Wegener's insight had come of age. The Earth was a dynamic planet with independent, rigid *lithospheric plates* of continental and oceanic rocks, floating over a mobile mantle of denser and hotter rock below. Plumes of magma welled up under the continents and created volcanic rifts that, in some cases, widened to form oceans, pushing the continents apart at a few centimetres a year. Where the continents collided, the less dense continental crust formed mountains while the denser ocean crust descended into the mantle along subduction zones, completing the cycle of convection. The Plate Tectonics Revolution swept the geological world in the late 1960s, and earlier theories were consigned to the landfill with remarkable speed.

As heat built up under the insulating bulk of Pangea, mantle plumes a hundred or a thousand kilometres across moved upwards.[9] A plume began to drive Africa and North America apart, generating lavas and intrusions—the Palisades Sill at New York, Cemetery Ridge at the Gettysburg Battlefield. The North Atlantic began to open some 200 million years ago in the Jurassic. By the end of the Cretaceous, the South Atlantic was opening between Africa and South America, the Indian Ocean between Africa and India, and the Southern Ocean between Australia and Antarctica. As Africa, Arabia, and India moved north to collide with Eurasia, nearly 30,000 km of mountain chains rose up from the Atlantic to the Pacific, closing off the 'lost ocean' of Tethys. The Andes and Cordillera grew where the Americas moved west over the Pacific Ocean crust.

As the continents drifted across stationary plumes and areas of hot convecting mantle, the land domed up hundreds of metres before tilting and lowering as

part of a process termed *dynamic topography*. Distant collisions and breakups stressed the Earth's crust and changed its elevation and, sensitive to water sources and slopes, the rivers responded over millions of years to this roller-coaster ride. Ancestral rivers continued to cross the continents, but young rivers came into being in youthful mountains. Few of our familiar rivers predate the breakup of Pangea.

How old is a river?

Unlike humans who are born at a moment in time, rivers have a timeless landscape heritage through which flowing water persists while rain and slope endure. American geoscientist Paul Potter set out to document the time when *ancestral rivers* began to follow their present courses, commenting: 'Water has always run down hill, has it not?'[10]

But establishing a river's genealogy is challenging.[11] Rivers may be drowned by a new lake or a rise in sea level, overrun by an ice sheet, lost in desert dunes, or blocked by lava and ash. Some rivers reoccupy valleys cut by ice or by rivers that long since moved elsewhere. Should we record the river's first appearance, or do such events reset the clock? Many continental-scale rivers consist of reaches with different ages, brought together when one river captured another. Is the river's age that of the individual reaches or the moment of connection?

And how do we know how old a river is? Unconformities and river strata may be dated using fossils or numerical methods. The dating of zircon grains in river sediment may provide clues to the age of the mountain belts from which they were eroded, and thermal analysis of mineral grains allows estimates of when mountains were rising and cooling. Seismic profiles and cores offshore may reveal delta sands and muds that could only have come from a large nearby river. The river courses that consumed the lives of adventurers may be examined in minutes using satellite images on a computer screen, where puzzling bends, waterfalls, and tributary junctions may reveal datable river events, and where relict river patterns can be discerned.

A picture is emerging of the history and age of the world's major rivers (Fig. 4.3), although many of the river histories set out here must be considered provisional. But the picture is one of rivers in constant flux as catchment areas expand and contract, drainage systems are captured

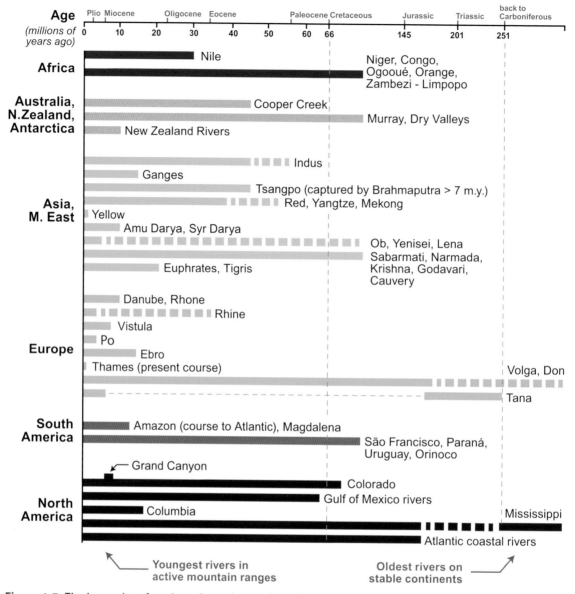

Figure 4.3 The longevity of modern rivers that are broadly in their ancestral courses.

by aggressive neighbours, rivers reverse their courses, and gorges the size of the Grand Canyon are carved in only a few million years. Some of the most elderly rivers live on drifting continents that have yet to collide, especially in Africa and Australia.

The rivers of Africa (Fig. 4.4) have had their share of drama. Few original African rivers still exist, although some flat landscapes may mark the long-dead surface of Gondwanaland. As Pangea began to break up, the Karoo and Paraná mantle plumes domed up southern Africa in the Jurassic and Cretaceous respectively, and river

systems that may have included the Zambezi, Limpopo, and Orange drained radially from the rising domes or were diverted round them.[12] Low-lying rifts drew the rivers into new paths to the Atlantic and Indian oceans, while the rise of rift margins that included the Guinea Highlands and Drakensberg Mountains generated new coastal and inland rivers. Since then, the continent has drifted over mantle plumes that have kept southern Africa unusually high as a 'superswell', and river capture in the Niger and Zambezi systems has generated angled bends, flow reversals, and waterfalls. But the pre-eminent

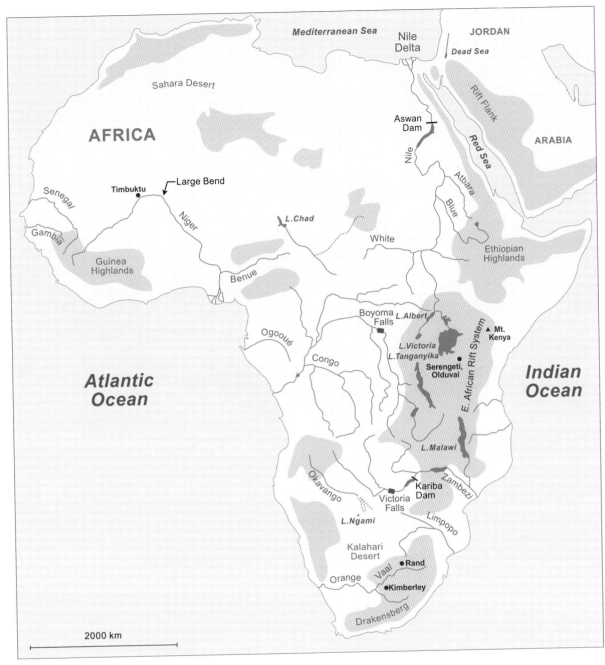

Figure 4.4 Modern rivers of Africa.

event for Africa's rivers was the rise of the East African Rift System.

Africa's rivers and East African Rift System

The East African Rift System, perhaps destined to widen into a future ocean, runs from the Red Sea down the eastern spine of Africa. A western branch that contains lakes Malawi, Tanganyika and Albert is nestled between rift flanks more than 5000 m high, and an eastern branch or Gregory Rift runs into the Ethiopian Highlands. Lake Victoria and the Serengeti Plains occupy a low area between the branches. The region boasts active volcanoes and the ancestral human cradle of Olduvai Gorge.

Figure 4.5 Sandbed rivers at Lake Manyara in the East African Rift Valley of Tanzania, with logs, baboons, and impala.

Foreshadowed by doming and lava eruptions from an enormous plume, the rift system began to form some 30 million years ago in the Oligocene, influencing many of Africa's big rivers.[13] The White Nile flows out of Lake Victoria and the Blue Nile and Atbara flow from the Ethiopian Highlands to a combined Nile that follows the Red Sea flank to the Mediterranean. The Congo flows west from the rift to the Atlantic and the Zambezi and Limpopo flow east to the Indian Ocean at the rift's southern terminus. Steep, 'short-headed' seasonal rivers bring gravel and sand from the uplifted flanks to the subsiding rift basins (Fig. 4.5), building alluvial fans into the lakes.

Detailed mapping of Africa's rivers goes back little more than 150 years.[14] Ptolemy's map from Roman times shows the Nile flowing through large lakes and the Niger traversing the continent from Ethiopia. Arab traders crossed the Sahara Desert and reached the Niger near Timbuktu, but no recorded adventurers penetrated further into the interior. Although navigators established the continent's outline, there were no inland seas accessible to large ships, and waterfalls blocked the way inland. In the early nineteenth century, the courses of the major rivers—the key to unlocking African geography—remained obscure to Europeans.

By 1880, a small group of adventurers had filled in the map of Africa, driven by colonial, religious, and commercial aspirations, big-game hunting, or simply an urge to discover. They were variously upstanding and odious, sometimes in rapid succession, and whether they intended it or not, European colonies followed hard on their heels. Nevertheless, their rapturous accounts provide a unique view of Africa's virtually pristine rivers. But the real explorers were the unnamed original inhabitants who walked across and out of Africa after Hominins evolved, familiar with every river bend.

The geology of East Africa was initially unravelled by British geologist Jack Gregory, for whom the Gregory Rift was named.[15] Gregory studied geology while working as a wool clerk, getting by with a few hours sleep a night.

A workaholic, he was once observed to entertain a child on one knee while correcting a manuscript proof with the other hand and talking enthusiastically to a friend. His interest in Africa stemmed from a childhood encounter with Robert Moffat, father-in-law of David Livingstone, who talked to him about Africa.

In 1892 Gregory joined an ill-fated expedition to East Africa where Britain had colonial ambitions. Seriously ill with dysentery, he barely made it back alive. Undeterred, he organized his own more mobile expedition. Used to little sleep, he guarded the camp at night and once thwarted a surprise attack by Masai warriors. Gregory discovered that the rift system was young and active, and the lack of folded rocks implied that the rift was widening with a faulted, downdropped centre and raised margins. He climbed alone across glaciers almost to the top of Mount Kenya, and he found stone implements under a lava flow, establishing their antiquity.

Gregory's insight into rift dynamics provided a framework for understanding the continent's river evolution. He returned to London in an emaciated condition, and as biographer Bernard Leake comments, he survived a risky venture through a combination of vigilance, astuteness, and sheer luck.

Nile River

The Nile fascinated Herodotus, variously considered the Father of History or the Father of Lies. Born shortly after 500 BCE in modern Turkey, he is widely considered to be the ancient world's first great critical historian for separating eyewitness accounts from hearsay. Although he may not have reached all the places that he claimed, his writings exhibit a cheerful zest for the juicy scuttlebutt of empire.

But it was the Nile—*the* river of the ancient western world—that caught his interest.[16] Flowing out of the desert, the Nile could not be sourced from melting snow, and unlike other rivers it flowed strongly in the hot summer. Herodotus thought that the Sun was driven from its normal course by stormy winter winds and drew up moisture in the south, drying the land but leading to summer rain. Returning to its former position in winter, the Sun drew water from the Nile, lowering the flow.

Roman expeditions to the source of the Nile bogged down in the wetlands of Sudan.[17] By the early seventeenth century, Jesuit priests had reached the source of the Blue Nile in Ethiopia, and in the 1760s Scotsman James Bruce reached a great cataract, barely visible for vapour. The Nile in this area was considered sacred, and his men were forced to cross the river barefoot. The people believe in nothing that you believe in, he was told, but only in the river.

In 1858, big-game hunter John Speke discovered a huge expanse of water, Lake Victoria, which he believed to be the source of the Nile. Exploring the river downstream from Lake Victoria on its way north through Lake Albert, he advised his men to shave their heads and bathe in the holy Nile where Moses had floated in his cradle. Samuel and Florence Baker reached Lake Albert in 1864 and, weak with fever, Baker rushed into the lake to immerse himself in the waters that fed the great river. The couple sailed down the lake in a canoe with a sail of Scottish plaid and found the point where the Nile entered from Lake Victoria. Further journeys ended after the canoe was lifted out of the water by a hippo.

The Nile mystery that had puzzled Herodotus was finally solved. Summer monsoon rains over the Blue Nile headwaters in Ethiopia bring floodwaters and sediment to Egypt, while year-round discharge from the White Nile through lakes Victoria and Albert maintains the river's base flow. For a river of its size, the Nile has a low discharge, traversing a desert and receiving virtually no tributaries downstream from the Atbara.

For all its fame as the world's longest river (nearly 7000 km long), the Nile is geologically young (Fig. 4.6).[18] Although it probably had antecedents, the first ancestral river along the modern course to the Mediterranean dates back about 30 million years, when the mantle plume raised the 'Roof of Africa'. Late in the Oligocene about 24 million years ago, the rising Red Sea rift flank forced the Nile westward through Libya as the Gilf River, and during the Miocene the river reversed its flow southwards into Sudan as the Qena River. Six million years ago during the Messinian stage of the Miocene, the Mediterranean was cut off from the Atlantic at Gibraltar and became an evaporating brine when its level fell more than 2000 m. In little more than half a million years, a reconstituted Nile was drawn in to resume a northerly flow, capturing the Qena and building a fan for several hundred kilometres across the salt deposits of the exposed sea floor, along with

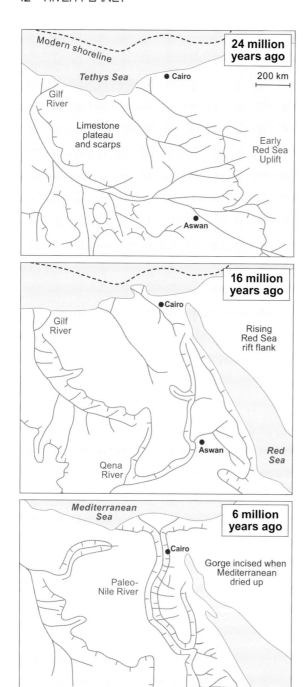

Figure 4.6 Evolution of the River Nile. Through the Oligocene and Miocene, drainage flowed northwest and southwest from the rising rift flank of the Red Sea, until the Nile settled into its northward course late in the Miocene. After Goudie, 2005.

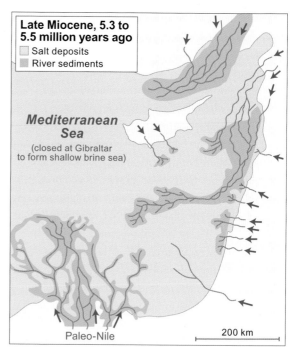

Figure 4.7 River systems in the Mediterranean basin during the Late Miocene (Messinian) desiccation and fall of sea level. After Madof et al., 2019.

other river systems (Fig. 4.7). Adjusting its gradient to the lowered sea level, the Nile carved a stupendous gorge more than 1000 km long and two kilometres deep, extending from the Mediterranean upstream to Aswan—a hidden wonder of the world and four times the length of the Grand Canyon. After the Atlantic reoccupied the Mediterranean, the flooded lower Nile filled the gorge with sediment.

The Nile has maintained its course since then. The cutting of the Blue Nile gorge in the Ethiopian Highlands, also rivalling the Grand Canyon in scale, reflects ongoing rise of the land over the past six million years. Lake Albert in the White Nile system had an endemic fauna until about half a million years ago, when Nile elements began to appear, and the fully connected White and Blue Nile system of today probably dates back only a few tens of thousands of years to the Pleistocene.

Niger River

The Niger follows Africa's most extraordinary river course. Rising near the coast, the river runs north into an inland delta before reforming through wind-blown dunes and swinging southeastward at the 'Large Bend' near Timbuktu. Picking up the Benue, the river reaches the coast after a journey of more than 4000 km.

At the corner of Africa where the Niger enters the Atlantic, three Cretaceous rifts met.[19] Two of the rifts were destined to widen as the Atlantic Ocean, but the third, the inland Benue Trough, became a *failed rift* later in the Cretaceous and has guided the Niger and Benue rivers to the ocean for the best part of 100 million years, fixing the offshore position of the Niger Delta, among the world's largest deltas. The Niger initially occupied the lower part of its present course (Fig. 4.8), whereas the larger Benue drained rifts in central Africa and parts of the Nile and Congo catchments until plumes domed up East Africa and the Hoggar and Tibesti massifs of the Sahara. Perhaps 15 million years ago the Niger captured the middle part of its course in the Iullemmeden Basin.

Since the early days of Atlantic rifting, rivers had flowed north from the Guinea Highlands into the Taoudenni Basin, while the Senegal River flowed west. As the Sahara Desert expanded, sand dunes blocked one of these rivers and a lake spillover at the Large Bend allowed the Niger to capture its upper course. The upper Niger expands into a large inland delta before narrowing through sand dunes into the Large Bend. A fully connected Niger may be only 10,000 years old.

The Niger's course was a matter of debate from the days of Herodotus and Ptolemy. In 1795 Scottish traveller Mungo Park barely survived a harrowing journey that took him up the Gambia and Senegal rivers until he came in sight of the Niger flowing in the bright morning light and as wide as the Thames in London.[20] Park's second expedition ten years later almost reached the sea. He canoed past Timbuktu and rounded the Large Bend until, far down the Niger, the expedition was mistaken for a war party and attacked. All hope lost, Park threw himself overboard and drowned. Finally, Richard Lander reached the sea in 1830, escorted by the canoes of a local ruler, the 'King of the Dark Water', and surviving lightning storms, cavorting hippos, and enormous crocodiles.

Congo River

In 1890, Captain Korzeniowski steamed up the Congo carrying a bag that contained a partly written novel.[21]

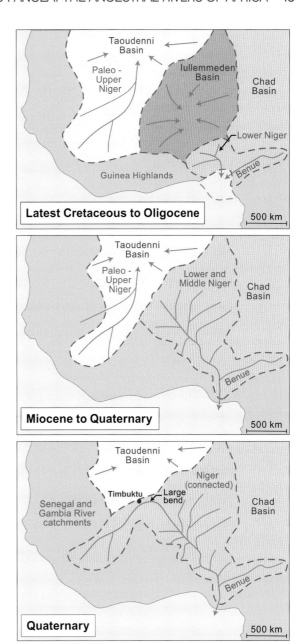

Figure 4.8 Evolution of the River Niger. During the Cretaceous, the Benue followed a failed rift and was the predominant drainage, until the Niger progressively captured its middle and upper reaches, forming a connected river perhaps as recently as 10,000 years ago. After Bonne, 2014.

Joseph Conrad—for it was he—had always wanted to visit Africa. When he was nine, he placed his finger on the blank central space in a map of the continent and decided with a resolve that he later lacked that he would go there.

Returning from Asia to London as a master mariner, Conrad saw a map of the Congo in a bookshop—'resembling an immense snake uncoiled, with its head in the sea, its body at rest curving afar over a vast country, and its tail lost in the depths of the land'.[22] With authorization from Belgian authorities, he walked inland through the jungle from the river mouth to circumvent Inga Falls, seeing firsthand the brutality of the colonial masters. Finding that his intended command had been wrecked a few days earlier, he became second officer on a steamboat which, over the next month, navigated 1500 km of the river to the foaming cataracts of Boyoma Falls, by far the world's largest set of cataracts by annual flow volume. Conrad's diary records the river's bends, woody snags, sand banks, and islands. Taking on board company agent Klein who died during the return journey, Conrad captained the boat back.

It was Conrad's only freshwater voyage. Denied command of a boat up a tributary river, which would probably have killed him, he returned to London deeply disillusioned with the colonial cruelties that he had observed. He now exchanged darkest Africa for darkest London. Fever, dysentery, and depression permanently damaged his health but, in his opinion, he was merely an animal before his Congo journey.[23] Often ill and house-bound, he went on to complete the novel *Almayer's Folly*, a story of colonial conflict in Borneo, and became a literary giant.

Second in flow volume only to the Amazon, the Congo occupies a broad lowland athwart the equator, which it crosses twice in its journey of more than 4000 km. The river maintains a strong year-round flow, always experiencing a wet season somewhere in its drainage basin, and it is incised for much of its length, fragmented by rapids and disconnected from the adjacent land. Early Portuguese navigators recorded muddy deluges from the river mouth that swept tree trunks and floating islands far out to sea, and the local people called the river *nzere* (Zaire), 'the river that swallows all others'.[24] By 1800, Europeans had journeyed little more than a hundred kilometres upstream, finding the Congo beset by rocky reaches, whirlpools, dense vegetation, and hordes of insects. Resolute and ruthless, Henry Morton Stanley traversed the Congo jungles in 1876 and was the first European to explore the Congo's primeval Great Forest, which he considered a region of horrors. On each occasion, he was accompanied by hundreds of expedition members, few of whom survived.

Steaming up the Congo less than fifteen years after Stanley's travels, Conrad encountered barbarous European traders who cruelly exploited the inhabitants. Everything became hateful, especially human beings. He reworked his experiences into *Heart of Darkness*, famous for the sinister brutality and moral turpitude of company agent Mr. Kurtz and his dying words 'The horror! The horror!' But Conrad's account brings the Congo dramatically to life:

> Going up that river was like travelling back to the earliest beginnings of the world, when vegetation rioted on the earth and the big trees were kings. An empty stream, a great silence, an impenetrable forest … The broadening waters flowed through a mob of wooded islands; you lost your way on that river as you would in a desert, and butted all day long against shoals, trying to find the channel, till you thought yourself bewitched … Trees, trees, millions of trees, massive, immense, running up high; and at their foot, hugging the bank against the stream, crept the little begrimed steamboat, like a sluggish beetle … The reaches opened before us and closed behind, as if the forest had stepped leisurely across the water to bar the way for our return … We were wanderers on prehistoric earth, on an earth that wore the aspect of an unknown planet…[25]

During the Cretaceous, the Congo flowed into the opening Atlantic near its present outlet, building up an offshore sediment fan, and the river may earlier have flowed out of Africa and across South America.[26] For a time thereafter, it may have flowed northwards to the Benue system and possibly as far north as Lake Chad, or it may have drained eastwards to the Indian Ocean. As the East African Rift Valley developed, the Congo Basin subsided, and rising coastal uplands are believed to have cut off the river from the sea, impounding an extensive lake. In its seaward reach, the river cuts steeply through a canyon and roars down Inga Falls with waves twelve metres high. This dramatic topography may have formed in the Pliocene when steep coastal rivers breached the lake, or the river may have maintained a steep, narrow course through rising coastal uplands. Some two million years ago in the Pleistocene, the Congo adopted its prominent sickle shape, influencing the distribution of our relatives the gorillas, chimpanzees, and bonobos. The river was in places a belt of channels and bars 10 km wide, and most primates could not readily make the crossing.

Ogooué River

North of the Congo is the Ogooué River of Gabon, made famous from the writings of adventurer and naturalist Mary Kingsley. Kingsley travelled with local people up the river by canoe in the 1890s, collecting fish and later climbing Mount Cameroon. In contrast to the writings of some vainglorious explorers, her account of the journey is amusing and self-deprecating:

> One appalling corner I shall not forget, for I had to jump at a rock wall, and hang on to it in a manner more befitting an insect than an insect-hunter … Some good souls helped the men haul, while I did my best to amuse the others by diving headlong from a large rock on to which I had elaborately climbed, into a thick clump of willow-leaved shrubs. They applauded my performance vociferously, and then assisted my efforts to extricate myself, and during the rest of my scramble they kept close to me, with keen competition for the front row, in hopes that I would do something like it again. … Round and round we spun in an exultant whirlpool, which, in a light-hearted maliciously joking way, hurled us tail first out of it into a current. … The shock being too much for M'bo and Pierre they were driven back on me, who got flattened on to the cargo of bundles … and the rest of the crew fell forward on to the bundles, me, and themselves … it's a wonder to me, considering the hurry, that we sorted ourselves out correctly with our own particular legs and arms.[27]

A delta corresponding to the ancestral Ogooué River lies offshore and dates to the Cretaceous opening of the South Atlantic. Thereafter, the river kept its position, pouring sediment into the ocean, and the Congo may at times have used this outlet.[28]

Zambezi River

Situated on the Zambezi, Victoria Falls (Mosi-oa-Tunya, *the smoke that thunders*) is the world's largest cascading falls, with a hundred-metre drop and a width of nearly two kilometres (Fig. 4.9)—a remarkable occurrence in an otherwise even landscape. In 1855 Scottish adventurer David Livingstone observed a vapour column like a burning grassland, visible far upstream. 'Have you smoke that sounds in your country?'[29] the inhabitants asked as they paddled down the Zambezi to the lip of the falls. Creeping awestruck to the edge, Livingstone looked down on an unbroken fleece of vapour that broke into 'myriads of small comets rushing on in one direction'. High overhead, the vapour rushed up like a steam jet, condensing to a smoke-like rain that ran down the dark rock, only to be swept up again by the vapour stream. Escaping fugitives, unable to arrest their flight, had been known to catapult over the lip onto the rocks below, while the pursuers, overcome with vertigo, crawled back from the abyss. Livingstone commented piously that 'scenes so lovely must have been gazed upon by angels in their flight', and he carved his initials on a nearby tree, 'the only instance in which I indulged in this piece of vanity'.

Livingstone had earlier crossed the waterless Kalahari Desert with oxcarts in search of legendary Lake Ngami near the Okavango River, hampered by the deaths of oxen from tsetse fly. Confused by mirages where the sun set over salt flats, he finally found the lake, a reedy expanse but a rushing torrent in the rainy season. He loved the Zambezi and the echoing roar of its lions. A later, less successful Zambezi expedition used a ponderous steamboat christened *The Asthmatic* for its heavy puffing.

Victoria Falls demonstrates the transience and dynamism of Africa's rivers.[30] As the Paraná dome raised adjoining parts of South America and Africa during the Cretaceous, the youthful Zambezi and Limpopo drained eastward to the Indian Ocean. The Limpopo catchment probably included the upper Zambezi and the Okavango, and the Zambezi and Limpopo occupy rifts associated with the earlier Karoo dome. *Barbed tributaries*, which join the main channel at a junction that faces upstream, suggest that parts of the Zambezi once flowed west to terminate as an *endorheic* (internally drained) system in the Kalahari Desert. Uplift inland may have separated the downstream Limpopo from its

Figure 4.9 Victoria Falls on the Zambezi River. Shutterstock©Vadim.Petrakov.

headwaters, rejuvenating the Zambezi. Both rivers built large deltas into the Indian Ocean from the Oligocene onwards.

In an act of piracy, the lower Zambezi first captured its middle reach in a low-lying rift and then during the Pleistocene its more elevated upper reach, generating a sharp change in gradient (a *knickpoint*) that became Victoria Falls. The falls retreated upstream at the rapid rate of some 20 km in 250,000 years, cutting Batoka Gorge with its angled bends into Karoo basalts of Jurassic age. As its sharp bends demonstrate, the Zambezi has only recently become a connected system, and it shares many fish and aquatic plants with the Limpopo. Sometime in the future, the dynamic river is expected to capture the Okavango, extending its catchment considerably to the west.

Tigerfish provide a surprising clue to the dynamic history of the Zambezi and other African rivers.[31] With a gaping jaw and formidable teeth, the fish weigh as much as a mastiff and hunt in packs, snatching swallows out of the air and attacking humans. Living in fast-flowing channels, tigerfish and their DNA are a guide to river connections since their first appearance in the Congo Basin ten million years ago in the Miocene. Similarity in DNA between Upper Zambezi and Okavango populations suggests that these now-isolated systems were once connected, as were the Nile and Niger populations across the Sahara. DNA analysis suggests that the fish migrated from the Niger into the Gambia and Senegal rivers, probably during a wet climatic period.

Orange River

In 1866 on the banks of the Orange River, a Boer farm boy noticed a flashing pebble and gave it to his sister for a plaything.[32] His mother passed it on to a neighbour who, unsure whether it was a diamond, sold it on cheaply until a local priest confirmed its identity by scratching his initials on a windowpane. But an expert declared that South Africa held no geological possibility for diamond fields, suggesting that the stone had been transported from elsewhere in the gizzard of an ostrich.

Two years later, a shepherd boy found another diamond, the Star of Africa, and traded it in for five hundred sheep, ten cows, and a horse. Prospectors rushed to the area, but the river sediments were unrewarding. Then, one evening in 1871, a group of dissolute prospectors in the uplands were roused by a servant who, thrown out for disorderly conduct, had dug up diamonds nearby. Soon ten thousand miners, mostly the scum of Europe, converged on the area. Troubled by the influx of greedy hopefuls, farm owner Johannes Nicholaas De Beer sold out to investors for a pittance. 'We have enough to buy twenty new wagons', said his wife, 'What have we to trouble about? We have enough.'[33]

The diggers assumed that the upland diamonds came from river terraces. Soon, however, they found diamonds in the solid rock below. An army of European and African miners excavated the 'Big Hole' at Kimberley, close to the original De Beer farm, climbing down ladders to the claims and raising dirt up wires that glinted in the moonlight. The mine remains the world's largest hand-dug excavation at 240 m deep. Underground workings took the mine down further until it closed in 1914 after yielding nearly three tons of diamonds.

The diamonds were in the top of a vertical pipe of *kimberlite* rock, formed by a Cretaceous eruption that brought up magma and diamonds from deep in the Earth. The pipes surfaced on the northern flank of the Drakensberg where the diamonds weathered out and rolled into the creeks. From there the Orange and Vaal rivers carried the diamonds westward where, over millions of years, they lodged among boulders in the channel and were preserved in terraces as the river cut down. Reaching the Atlantic, the diamonds drifted along the beaches or were swept offshore. In 1908 a railway worker in Namibia found a diamond stuck to the oil on his shovel, leading to another diamond rush.

With its headwaters in the rift flank of the Drakensberg, the Orange was present during the breakup of Gondwanaland during the Jurassic and Cretaceous, flowing west from the Karoo dome and possibly crossing into South America before the Atlantic opened.[34] The river has also responded to the uplift and tilting of southern Africa as it drifts over a dynamic mantle. For at least 100 million years, the Orange has followed the same general course, carrying its precious cargo of diamonds to the sea.

African rivers of the future

It is likely that Africa will continue to collide with Europe, closing off the last remnant of the Tethys Ocean and forming the Mediterranean Mountains. Somalia in the Horn of Africa will break away from Nubia to the west along a widening East African Rift System. A broadening Atlantic will convey North America against Asia. This will be *Amasia* 250 million years from now, an American–Asian coalition.[35]

From the Mediterranean mountains, rivers will flow south across Africa. The *Herodotus River* (I might guess) will reverse the Nile, which will flow out through

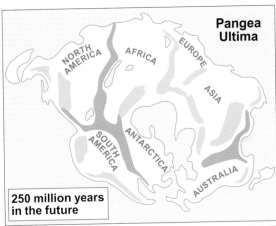

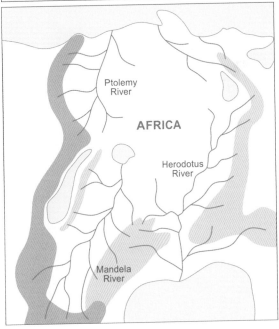

Figure 4.10 How African river systems might look in 250 million years' time in the Pangea Ultima reconstruction. After Williams and Nield, 2007.

the Congo into the South Atlantic. The *Ptolemy River* will run south through the Sahara to join the Niger and Benue. As the rivers carve gorges through the mountains, tigerfish will invade the rivers of France and gorillas will move through the jungles into Italy. And the *Zambezi* and *Orange* will continue in their accustomed courses until they are 350 million years old.

Unless the Atlantic closes once more to form the Afro-centric supercontinent of *Pangea Ultima*, thrusting up mountain chains where the Americas meet Africa (Fig. 4.10). Hemmed in by mountains, the Herodotus River will flow south to the remnants of the Indian Ocean, picking up the Zambezi. The Ptolemy River will flow out where the world ocean touches northwest Africa, picking up a westward-flowing Niger. And where the Orange once ran west, the majestic *Mandela River* will run inland from the Drakensberg, picking up a reversed Congo and Ogooué blocked by mountains to the west and flowing out into the heart of Africa.

CHAPTER 5

HOT AND COLD: THE RIVER HISTORIES OF AUSTRALIA, NEW ZEALAND, AND ANTARCTICA

Rivers of Australia

With the 1901 Christmas exams over at the University of Melbourne, Jack Gregory set out to circumnavigate the salt flats of Lake Eyre or Kati Thanda (Fig. 5.1).[1] He alighted at the railhead and camel centre of Marree (then Hergott Springs) at the southern end of the Queensland Road cattle trail that followed the waterholes through the Channel Country. To select suitable camels, Gregory consulted the War Office Manual on Transport but found it unhelpful: camels could not work in the hot sun, and the traveller should avoid animals with broken jaws and poorly attached legs. The expedition included five students, Aboriginal guides, a cook, a camel driver, and nine camels—the sweet-tempered 'eight saints' and 'Fireworks'. The camels were prone to spit their cud on the travellers when overloaded.

Gregory wanted to find the bones of gigantic extinct marsupials, which had been reported from the river sands near Kati Thanda. But he was also an adventurer who revelled in *terra incognita*, as well as a humorous writer with 'ink in the blood'[2] and much of the publicist in his nature. In the intense heat of the austral summer, Gregory had five weeks to round the lake and catch the train back from Warrina for the start of the next semester.

By this time, the European mapping of Australia had been underway for more than a century. After Captain Cook mapped the southeast coast in 1770, a penal colony was established at Botany Bay near Sydney, apparently the only use that an unimaginative colonial power could envisage for Australia. It was not until 1813 that adventurers crossed the Great Dividing Range, following ridges rather than the valleys, which ended in towering cliffs.

Did the rivers flow to an inland sea west of the Dividing Range in the unmapped 'ghastly blank'?[3] In 1828, Charles Sturt followed a west-flowing stream that he named for Governor Darling, and the next year he sailed down the Murrumbidgee to the Murray River which, he found, flowed to the Southern Ocean. On a later expedition, Sturt observed Kati Thanda north of Mount Hopeless and set off up the Murray and Darling rivers past a litter of iron ore (later the mining district of Broken Hill). In the blistering summer heat, they stuck at a waterhole, unable to go forward or back as the temperature rose to 55°C in the shade. Screws dropped out of boxes, lead from pencils, and matches ignited on the ground. When the rains returned, they journeyed on up Strzelecki Creek, crossing sand dunes and creeks snow-white with salt, before turning back.

Still hoping for an inland sea, Sturt set out again and this time found a magnificent sheet of water with belts of flooded trees. Eastward the river broke up into a myriad anastomosing channels (Fig. 5.2), with only the stony desert beyond. The Darling River had ceased to flow but, following the waterholes, they struggled into Adelaide barely alive. Sturt named the anastomosing channel Cooper's Creek (now Cooper Creek), reluctant to call it a river because it had little flow.

In 1860, the Burke and Wills expedition left Melbourne, cheered on by a huge crowd. Robert O'Hara Burke was a police officer and former cavalry soldier who scrawled poems in several languages on the walls of his prison station. William John Wills was a young surveyor whose father had hurried to Australia for the gold rush. Wills hated the wild gold diggings but loved the bush country. With camels brought from India, they established a depot at Cooper Creek by a waterhole with an impressive coolibah tree, close to modern Innamincka. Burke and Wills with two companions travelled north through the summer heat to the Diamantina River near present-day Birdsville, following the Flinders River to the Gulf of Carpentaria.

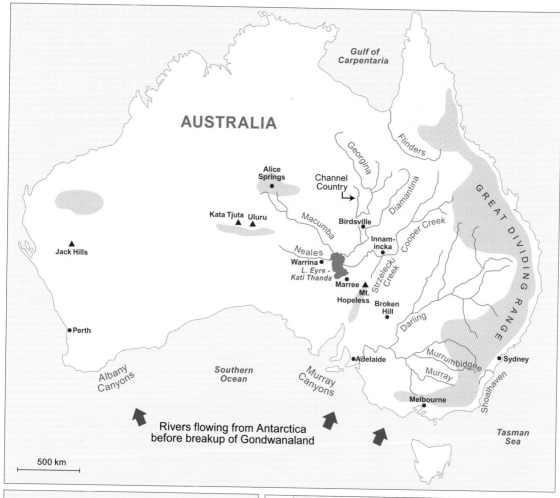

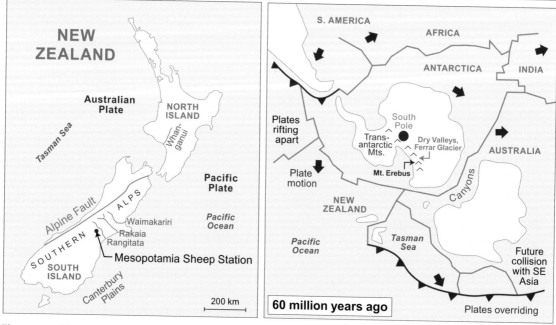

Figure 5.1 Modern rivers of Australia and New Zealand, with part of East Gondwana 60 million years ago. Tectonic map after Seton et al., 2012.

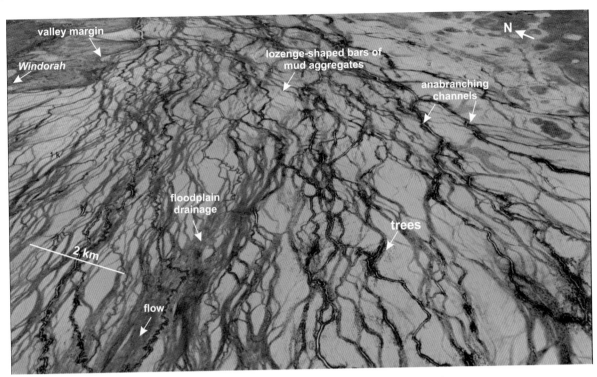

Figure 5.2 Anastomosing channels of Cooper Creek cut into tough floodplain mud, near Windorah in central Australia. Maps data: Google, CNES / Airbus Maxar Technologies.

The return journey was a nightmare. Reaching the Cooper Creek depot at the limits of their strength, they found a fresh mark on the coolibah tree and a buried box with a note. The rest of the party had left earlier on the day of their arrival after waiting well beyond the appointed time. Too weak to reach Mount Hopeless, they staggered back to the depot where John King, the only survivor, was saved by the Aboriginal community.

Search parties for Burke and Wills revealed many of the region's mysteries. Peter Egerton-Warburton described Kati Thanda as

terrible in its death-like stillness and the vast expanse of its unbroken sterility. The weary wanderer, who, when in want of water, should unexpectedly reach its shores, might turn away with a shudder from a scene which shut out all hope—he could hide his head in the sand-hills, and meet his fate with calmness and resignation; but to set his foot on Lake Eyre would be like cutting himself off from the common lot of human beings.[4]

But soon, settlers were moving in and, by Gregory's time, few Aboriginal inhabitants remained. The party shrugged off the summer heat and the dreadful experiences of Sturt, Burke, and Wills. The settlers at Hergott Springs were incredulous: Gregory, newly arrived from England, was either unaware that the austral summer was at Christmas or he was crazy. A few years before, 5000 head of cattle had been overcome by a summer dust storm on the Queensland Road, suffocating or fatally bogged in a stampede to a salt pool.

The camel party reached Cooper Creek, now a dry expanse of mud among sand dunes, with a high floodline of stranded wood. Moving through a dust haze that Gregory likened to a Scotch mist, they put up a flock of cranes or brolgas, the 'native-companions' that dance at dawn in an elaborate ritual. They made tea from hot water pouring from a deep water bore to supply cattle in the Great Artesian Basin. 'Anyone who wants it hotter is greedy,'[5] said the last remaining stockman.

Near Kati Thanda the creek bed and bushes glittered with salt around brine pools (Fig. 5.3A). Their Aboriginal guides found fresh water by following dingo tracks or by digging around the pools, where clay trapped rainwater trickling through the sand. Eroding from ancient river strata in the banks and littering the dry river-

Figure 5.3A Cooper Creek near Kati Thanda (Lake Eyre), where the river is reduced to saline ponds with encrusted gypsum.

Figure 5.3B Dry, low-season anastomosing channel of the Georgina River, Queensland, cut into tough floodplain mud. The channel is relatively stable, with minor bank erosion that has caused trees to topple. Sand-sized quartz and mud aggregates form ripples and low dunes on the channel floor and on the bench at left, with mud drapes on steeper banks.

bed, then as now, were the fossilized bones of kangaroo, crocodile, and a species of lungfish that was later named for Gregory. At last they reached the shimmering salt crust of Kati Thanda, and Gregory took a 'short cruise out over the lake'.

Crossing the sand dunes, they reached the Warburton River below the confluence of the Diamantina and Georgina system (Fig. 5.3B), alive with pelicans and ducks. It took them three hours to dig out a camel that bogged down, and a dust storm sent thousands of cockatoos screaming down the river ahead of the blast. Once more the riverbed was strewn with fragments, this time from gigantic extinct wombats. Gregory envisaged herds of wombats and kangaroos flourishing around a

large lake with crocodiles and lungfish. When did the lake dry up? And how had the rivers evolved?

The final days of the expedition were a race against time, through trackless country near the Macumba and Neales rivers and in sight of mountains west of the lake. They rode their camels triumphantly into Warrina station with seven hours to spare. A member of the expedition welcomed them with a song that seemed to combine the Melbourne University Anthem with the Hallelujah Chorus.

Much less demanding than his gruelling African safaris, Gregory's expedition was well publicized. Nevertheless, his exploits would have been forgotten had he not, with his customary panache, published in 1906 *The Dead Heart of Australia*, a phrase that became firmly entrenched. In 1973, Osmar White's childrens' book *The Super-Roo of Mungalongaloo* featured Dr Alastair Angus Archibald McGurk who travels by camel to the 'dead centre of the Deadibone Desert'. Gregory had been immortalized.

Australia is the lowest and most arid continent with Kati Thanda, the world's largest ephemeral lake, below sea level in a huge inland drainage basin.[6] The continent is positioned within the subtropical high-pressure belt where sinking hot air limits the rise of moist air that might cause condensation and rain, although tropical cyclones bring rain to northern areas. The trade winds blow across the Tasman Sea from the southeast, laying down a belt of rain on the Great Dividing Range along the east coast, with a rain shadow to the west. Rain is greatly reduced across eastern Australia during El Niño years, linked to atmospheric pressure and sea-surface temperatures in the Pacific Ocean. Not surprisingly, with less predictable rainfall and stream flow than any other continent, Australia is a land of ephemeral rivers that may carry more flow in a week than over the next year. The Diamantina may be the world's most variable river. Unlike their tropical and temperate counterparts, many rivers dwindle in volume downstream, and three-quarters of Cooper Creek flow evaporates or trickles into the sand.

Like Africa, Australia is a drifting continent far from plate boundaries and built around ancient cratons. Some have considered it an inactive continent still crossed by ancestral rivers that once traversed a unified Gondwanaland. But everything changed during the Cretaceous, when Australia drifted away from India and Antarctica and the Tasman Sea began to open between Australia and New Zealand about 100 million years ago, with a rising rift flank along the Great Dividing Range.[7] Thereafter Australia was the fastest-drifting continent, moving more than 3000 km north in less than 50 million years. For much of that time, northern Australia has been subsiding as it collides with Southeast Asia while southern Australia has been rising. The continent is tilting and deforming, buffeted by conflicting forces—rifting away from Antarctica, in collision with New Guinea, and affected by the distant rising Himalaya. There are young volcanic craters and lavas that flowed down river valleys.

The Australian landscape can be deceiving. Land surfaces and river valleys may appear to be relics from Gondwanaland or earlier supercontinents, but old landscapes have been buried repeatedly and then, over millions of years, slowly uplifted and exposed again to leave a landscape with a false appearance of great age. In contrast, the resistant *regoliths* of calcrete, silcrete, and ferricrete (cemented soils developed on bedrock), which cap hills across the interior, contain plant fossils as old as Cretaceous in places: they appear to be young but are of considerable antiquity. The rivers tell a subtle story, centred around the dynamic topography of the drifting continent and the history of the Great Dividing Range.

Rivers of the Great Dividing Range

The Great Dividing Range is a magical mountain belt, 3500 km long, where the remote valleys contain conifers until recently known only as fossils. The jungle is a haunt for poisonous snakes, spiders, and leeches. But in the valleys at dawn, the clear song of the bell bird rings out from the eucalyptus trees and duck-billed platypus dive for crayfish in the creeks. The forests echo to the laughter of the kookaburra and the riverbanks are drilled with wombat burrows. At night, a flashlight beam reflects the glowing eyes of nocturnal marsupials in the branches. The rivers, too, are remarkable. They may receive a metre of rain in a single day or flow so weakly that storm waves block their coastal outlets with sand. And in times of drought the air is suffused with a grey pall as the desert dust sweeps out to sea.

The Dividing Range has a long history as a mountain belt. Behind the Sydney Opera House and near Botany Bay—many a convict's first view of their new home—

Figure 5.4 Hawkesbury Sandstone of Triassic age, near Sydney, Australia, laid down by rivers flowing from Antarctica before East Gondwana broke up. Braided-river sandstones at cliff base and top, with sandbar with accretion surfaces, are separated by a large abandoned channel filled with mudstone. People for scale (circled).

stand the outcrops of the Hawkesbury Sandstone, laid down by one of the largest braided rivers known in the geological record.[8] The Hawkesbury river emerged from the ancestral Dividing Range with channels up to 20 m deep filled with fossilized sand bars that are still visible in the cliffs (Fig. 5.4). Flowing during the Triassic when Gondwanaland was still intact, the river's headwaters lay in an Antarctic mountain range (now under the icecap) from which the drainage radiated out across Australia, Africa and India—continents now separated by thousands of kilometres of young ocean.[9]

More recently, the Great Dividing Range has been a model of unusual geological stability.[10] Much of the range formed late in the Paleozoic during the assembly of Pangea, and before erosion cut them down, the mountains would have been of great height. Remarkably, the continental divide between rivers that flow to the coast and the interior has been pinned in about the same location for the past 180 million years, long before the Tasman Sea began to open. And the rivers have cut their valleys surprisingly slowly. About 30 million years ago in the Oligocene, the Shoalhaven River broke through a dam of lava that had flowed down its valley and impounded a lake. The river carved out an impressive gorge through the rocks below the lava, the age of which provides a time marker for the start of incision. However, the river cut down at a rate of only some ten metres in a million years, and it will take the river forever to carve a fretwork of canyons like those of the Colorado Plateau, if it ever does.

Rivers of the Red Centre

Jack Gregory's dead heart of Australia is a landscape like no other. In the humid north along the Gulf of Carpentaria, the monsoon lays down 1000 mm of rain a year, with the rivers reduced to yellow sand threads among groves of trees during the dry season (Fig. 3.1B). Southward down the arid droving roads of Gregory's day, life depends on the Channel Country waterholes, opaque with river clay and windblown dust, set in a brilliant landscape—the uncanny red of dunes and weathered rock, the yellow of savannah, the glitter of salt pans. There are the scuttling tracks of lizards, termite mounds like model medieval spires, and the clay 'chimneys' where crayfish have burrowed in the creek beds. Stockmen on horseback once moved the cattle down the Birdsville Track to waterholes (billabongs) alive at dusk with the calls of corellas and budgerigars, sleeping out for months at a stretch in their bedding rolls or 'swags', as recorded in Banjo Paterson's song *Waltzing Matilda*. The flood markers stand barely shy of cattle stations on the highest ground, but in El Niño years, the stockmen marshal skeletal herds to the last vestiges of grass.

The Red Centre has a geological history typical of cratons, with a record of slow subsidence and erosion, unconformities, and broad folds.[11] The thin and patchy strata have made it difficult to document the river histories. From the Great Dividing Range, rivers carried a flood of sediment westward through much of the Mesozoic. As Australia drifted away from India in the Cretaceous and its northern margin subsided, the Tethys Ocean flooded much of the continent and dinosaurs trampled out footprints along rivers that flowed north to the ocean. Gentle doming of the low-gradient interior during the Eocene some 50 million years ago turned an ancestral Cooper Creek southward into the Red Centre. As the climate became more arid, alkaline lakes formed in the Oligocene and Miocene, some of them within the river channels as in Cooper Creek today (Fig. 5.3A).

Near the start of the Pleistocene, the climate became more strongly arid, and dunes and stony deserts began to form. Since then, oscillations of climate and precipitation have greatly affected the rivers. Aerial photos of Cooper Creek show traces of sandy point and scroll bars some 75,000 years old, when a big meandering river brought more than five times the discharge of the modern creek to Lake Eyre, which has fluctuated in depth accordingly.

As Australia and Antarctica rifted apart some 160 million years ago, other Australian rivers flowed south to the low-lying, widening rift. By the end of the Cretaceous, the Murray-Darling drainage system was flowing west and south from the Great Dividing Range much as it does today. Off the mouth of the Murray River and south of Perth, some of the world's deepest submarine canyons, two kilometres deep, convey river sediment to the Southern Ocean, as the ancestral rivers and canyons did 100 million years ago.[12]

And, far to the west near Alice Springs, the red rock masses of Uluru (Ayers Rock) and Kata Tjuta (The Olgas) rear up above the desert. Scarred by lightning strikes and swept by waterfalls during heavy rain, they have been sacred sites for Aboriginal people for thousands of years. For at least 70 million years since the Cretaceous and perhaps long before, the rock masses have been rising from an eroding plain,[13] a matter of wonder to passing dinosaurs and gigantic marsupials.

Jack Gregory puzzled over the bones of gigantic marsupials from the rivers around Kati Thanda. When were the animals living, and did the first humans see them? The megafauna included the three-ton giant wombat *Diprotodon* (the largest marsupial ever to shamble across Australia), a massive marsupial lion, the 'super-roo' *Procoptodon*, and the flightless emu-like bird *Genyornis*, with massive legs and as tall as a human.[14] The megafauna became extinct more than 40,000 years ago: could climate change—so effective in changing the rivers—have caused their decline? But the rivers and lakes were high at that time, and it is more likely that our species, newly arrived on the continent some 65,000 years ago, hunted them down or drove them to extinction by burning off the vegetation. Depicted in rock art across Australia are Aboriginal images that may show these lost Pleistocene creatures, including a compelling ochre image of *Genyornis*. The first Australians reconstructed the Pleistocene world as lovingly as any geologist.

Rivers of Antarctica

Robert Falcon Scott and seamen Edgar Evans and William Lashly manhandled their sledge down the Ferrar Glacier in Antarctica, heading back to their ship *Discovery*.[15] Reduced to meagre rations they dreamed of food—Devonshire cream for Scott, pork for Evans, and apples for Lashly. It had been a gruelling journey through the austral summer of 1903. Forging a route up the rugged glaciers that crossed the Transantarctic Mountains to a supply depot, they found that a gale had whirled away the Royal Geographical Society's *Hints to Travellers* with its tables for navigating by the stars. They emerged onto a high ice plateau and pressed on for several weeks through blizzards. At night in the frigid tent, Scott gained solace from Charles Darwin's accounts from the *Beagle*. With provisions dangerously low they turned back.

On the steep descent to the Ferrar Glacier, Lashly lost his footing and they tumbled a hundred metres, bruised but uninjured. Scrambling to their feet, they found that they had reached the glacier entrance with the smoking volcano of Mount Erebus beyond. Then Scott and Evans plunged suddenly down a crevasse where they dangled from their harnesses as the sledge teetered on the brink. At last they reached a vital depot where Lashly 'sang a merry stave as he stirred the pot'.

Then, to their surprise, the glacier ended abruptly at a muddy moraine, 'a splendid place for growing spuds' said Lashly. Beyond stretched a valley with sandy streams (Fig. 5.5), and late on Christmas Eve they saw the masts of *Discovery* in the distance. Scott was fascinated by this unexpected discovery:

> as we ran the comparatively warm sand through our fingers and quenched our thirst at the stream, it seemed almost impossible that we could be within a hundred miles of the terrible conditions

Figure 5.5 Taylor Valley, Antarctica, a dry valley that was cut by rivers about 55 million years ago and subsequently overrun and deepened by ice. Shutterstock©DaleLornaJacobsen.

we had experienced on the summit … Below lay the sandy stretches and confused boulder heaps of the valley floor, with here and there the gleaming white surface of a frozen lake and elsewhere the silver threads of the running water; far above us towered the weather-worn, snow-splashed mountain peaks … I cannot but think that this valley is a very wonderful place. We have seen to-day all the indications of colossal ice action and considerable water action, and yet neither of these agents is now at work. … It is certainly a valley of the dead; even the great glacier which once pushed through it has withered away.[16]

Over the next decade, Australian scientists studied the Dry Valley. As part of Ernest Shackleton's 1907–9 expedition, T.W. Edgeworth David, the first trained geologist to visit Antarctica, noted thick river gravels and valley margins cut through by glaciers. Griffith Taylor, a student of David, mapped the valley, which Scott named after him, panning in the streams but finding no bright swirl of gold.[17] Apsley Cherry-Garrard, a fellow expedition member, penned a memorable description of Taylor:

Old Griff on a sledge journey might have note-books protruding from every pocket, and hung about his person, a sundial, a prismatic compass, a sheath knife, a pair of binoculars, a geological hammer, chronometer, pedometer, camera, aner-oid and other items of surveying gear, as well as his goggles and mitts. And in his hand might be an ice-axe which he used as he went along to the possible advancement of science, but the certain disorganization of his companions.[18]

At a service where Scott read the psalm verse 'the little hills skipped like rams', Taylor muttered, 'Damn nonsense! The psalmist was no geologist,'[19] receiving a glare from Scott.

From the first discovery, opinion was divided about glaciers or rivers as valley sculptors.[20] Scott and Taylor seem to have presumed a glacial origin: the valley walls were planed off, suggesting ice action, and glacial debris covered the floor. On the other hand, David's observation of thick gravels suggested a contribution from meltwater, and later geologists thought that glaciers had occupied a former river valley. Recent studies of the Transantarctic Mountains show that the dry valleys were originally river valleys, carved out about 55 million years ago in the Eocene as Antarctica split from Australia, with sinuous patterns and long profiles graded to a former ocean level. The mountain chain is the uplifted flank of a rift system that the rivers transected, and the valleys were once close to the Murray River canyons, both sets of rivers probably feeding into the same developing rift but now located on opposite sides of the Southern Ocean. After Antarctic ice began to form more than 30 million years ago, glaciers further deepened the valleys until ice sheets overrode the area with their active ice streams, large subglacial lakes such as Lake Vostok, and opportunities for valley sculpturing by outburst floods of meltwater. The valleys are relict river landforms, weathering slowly in the cold Antarctic desert.

Rivers of New Zealand

The Canterbury Plains of New Zealand extend eastward from the snow-capped peaks of the Southern Alps, the country of *Erewhon*. In his 1872 eponymous satire, Samuel Butler placed his dystopian society in a country so remote that, reversed, it is truly Nowhere. Deep in the mountains, Butler's hero encounters a society that outlaws machines and considers illness a criminal offence. One by one Erewhon's institutions (and, by extension, Victorian society) are exposed for their shortcomings. Affairs of state, religious traditions, the legal system, and the Colleges of Unreason that provide a treatment termed education—none are exempt. New Zealand's rivers buttress the curious social landscapes of Erewhon: 'Property, marriage, the law; as the bed to the river, so rule and convention to the instinct; and woe to him who tampers with the banks while the flood is flowing.'[21]

Erewhon lies in one of the Earth's most unstable terrains.[22] The Alpine Fault runs down the South Island, separating the Australian and Pacific plates, and some areas have risen at nearly a centimetre a year—a rate that, if sustained, would raise the rocks from 10 km depth to the surface in as little as a million years. Wide braided rivers emerge from mountain gorges as roaring glacier-fed torrents under snowmelt and cloudbursts but placid in late summer—the Rangitata, Waimakariri, and Rakaia rivers (Fig. 5.6), named by Polynesian

Figure 5.6 Rakaia River, a braided gravel river emerging from the Southern Alps onto the Canterbury Plains of New Zealand.

adventurers who settled the islands after the thirteenth century CE. As the sun sets, long shadows slant across the plains and curious dappled patterns appear that mark former channel courses, picked out by subtle changes in vegetation. The farms host many of New Zealand's 30 million sheep, which could so easily outwit the small human population if only they knew how.

After the continent of Zealandia broke away from Australia and Antarctica in the Cretaceous about 80 million years ago, much of the landmass lay under the ocean until the Miocene some 10 million years ago, when a small part emerged to form modern New Zealand.[23] Braided rivers poured gravel and sand from the rising mountains onto the ancestral Canterbury Plains, filling lakes dammed by tectonic uplifts. The river deposits were eroded and recycled in their turn as faults brought them to the surface, along with dense gold particles that were progressively concentrated by the river flows. Marine-migratory fish entered river catchments that were periodically isolated by earth

movements, evolving new freshwater species in as little as half a million years in parallel with the tectonic events. Taking advantage of tectonic events and river capture, some species have spread through the coastal rivers and have even crossed the drainage divide of the Southern Alps.

In conflict with his father, Butler the prodigal son emigrated to New Zealand in 1860 and found a sheep station named Mesopotamia in the Rangitata Valley, situated like the ancient cities of Iraq between river channels.[24] For five years before returning to England, Butler enjoyed the tough life, existing on mutton, bread, and tea until the station was established. He found the sheep to be manageable, but the bullocks that drew his cart were not, and he declared that the word 'bullocks' would be found written on his heart when he died.

The unpredictable Rangitata River dominated Butler's life. Adventurers had drowned in the wild torrent, but Butler forged a path up the gloomy and desolate valley. Landforms were named for him—Mount Butler, Erewhon Col. His description of the rivers that led to Erewhon is compelling:

> The river-bed was here about a mile and a half broad and entirely covered with shingle over which the river ran in many winding channels, looking, when seen from above, like a tangled skein of ribbon, and glistening in the sun. We knew that it was liable to very sudden and heavy freshets; but even had we not known it, we could have seen it by the snags of trees, which must have been carried long distances, and by the mass of vegetable and mineral *débris* which was banked against their lower side, showing that at times the whole river-bed must be covered with a roaring torrent many feet in depth and of ungovernable fury ... The gorge was narrow and precipitous; the river was now only a few yards wide, and roared and thundered against rocks of many tons in weight ... I could hear the smaller stones knocking against each other under the rage of the water ...[25]

Erewhon rose out of Butler's pioneering struggle and the perspective that it gave him on a civilization that paled before the New Zealand wilderness: 'I am there now as I write; I fancy that I can see the downs, the huts, the plain, and the river-bed—that torrent pathway of desolation, with its distant roar of waters. Oh, wonderful! wonderful! so lonely and so solemn.'[26]

Others found the landscape less enthralling. Growing up on a South Island farm, nuclear physicist Ernest Rutherford, the originator of radioactive dating, applied for a scholarship to Cambridge University. He was digging potatoes when the mailman appeared with a telegram: he had won the scholarship. Throwing down the spade, Rutherford declared that he would never dig another potato again.

CHAPTER 6

YOUNG AND RESTLESS: THE EVOLVING RIVERS OF ASIA

The water tower of Asia

Nowhere have I felt the life of rivers more powerfully than here at the top of the world. Starting the ascent before dawn, we climb through snowfields to the Thorong La pass with its cairn and prayer flags at 5400 m (nearly 18,000 feet) above sea level. The Himalayan peaks and the Tibetan Plateau form an astounding vista, an immensity of space and geological time at once exhilarating and profoundly diminishing. Ice avalanches rumble down and the brittle air feels perilously thin. Vultures quarter the lower slopes, and I remember John Muir on an Alaskan glacier shouting to the circling ravens 'Not yet, not yet'.[1]

Far below, the Kali Gandaki River winds its way south to the Ganga (Ganges) (Fig. 6.1A). Dwarfing the Grand Canyon, the river gorge is more than five kilometres deep, cut through the towering peaks of Annapurna and Dhaulagiri where, in the early nineteenth century, surveyors first proved that the Himalaya was the highest range on Earth. The river follows fault lines where scalding water gushes up between the boulders and trickles from rock clefts. Rock masses bigger than houses mark enormous landslides, and a metal bridge that I crossed some years before dangles pathetically down a rockface. Along the route to the pass, caravans of mules and yaks follow dangerous paths along the valley walls, as once they followed the Silk Roads from China.

Pilgrims head for the monastery of Muktinath where a spiral ammonite sits on a temple altar, a Jurassic fossil from the lost Tethys Ocean and an image of the wheel of life. Exposed along the Kali Gandaki valley are cliffs of Cretaceous strata laid down by rivers that once flowed north to Tethys (Fig. 6.1B).[2] The rivers reversed their flow after India collided with Asia and Ordovician rocks with marine fossils were thrust to the top of Qomolangma (Mount Everest).[3]

Figure 6.1A Kali Gandaki Valley, more than 5 km deep, running south between the 8000 m peaks of Annapurna and Dhaulagiri in Nepal. The river was antecedent to the most recent period of tectonic uplift, cutting down to keep pace with the rising mountains. Note talus cone at centre right and large landslide blocks in the river.

Figure 6.1B Cretaceous sandstones in a cliff along the Kali Gandaki River at Kagbeni, Nepal. Palaeoflow from cross-beds in the cliff was north across India prior to collision with Asia, whereas the modern river flows south towards the viewer. Note Quaternary terraces at lower right, sculptured into agricultural terraces with crops. Pale flat-lying clays in lower cliff at centre are the deposits of a lake dammed by landslides.

I am in the heart of the 'water tower of Asia', a stupendous upland of radiating rivers (Fig. 6.2). A suite of large rivers traverses Southeast Asia, and the Yangtze and Huang He (Yellow) cross China to the Pacific, the latter engineered by Chinese emperors over millennia (see Chapter 16). The Lena, Yenisei, and Ob flow north to the Arctic Ocean, and the Amu Darya and Syr Darya flow west into what remains of the Aral Sea, deprived of water by decades of irrigation. The Yarlung Tsangpo runs along the grain of the mountains and swings south to become the Brahmaputra, a connection that was in doubt until the late nineteenth century. South of the mountains, the Ganga picks up the Yamuna, which flows past the marble tomb of the Taj Mahal, and curves eastward to meet the Brahmaputra as the united Padma River. Garlanded corpses, reduced to ashes in cremation

fires, float down the sacred river to the sea, and an occasional skull washes up along the cliffs.

To the west runs the Indus, and somewhere between the Indus and Ganga headwaters lies the lost Saraswati River, the mother of all rivers in the ancient text of the *Rigveda* (see Chapter 15). Across Peninsular India, monsoonal rivers head east to the Bay of Bengal, and closer at hand the Luni passes the fortress of Jodhpur where the raja's harem once peered through a stone fretwork to the domes of far-off palaces. And on the bank of the Sabarmati stands Mahatma Gandhi's ashram and his room with a few possessions—a sleeping mat, a spinning wheel, a bookshelf, his sandals, and his iconic round spectacles.

My first sight of an Indian river was the celebrated Ganga with its dry-season expanse of sand bars. Returning to Delhi at dusk, the Sikh driver roared

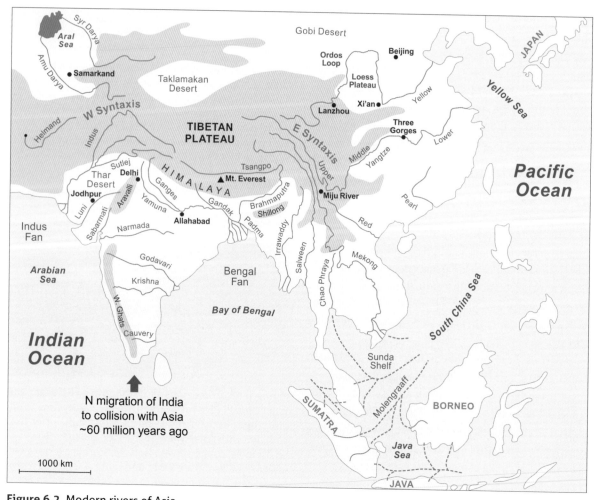

Figure 6.2 Modern rivers of Asia.

YOUNG AND RESTLESS: THE EVOLVING RIVERS OF ASIA   61

past trucks and oxcarts on the crowded road, swerving against the oncoming headlights. The passengers—Hindu, Jain, Muslim, and Christian—debated whether one should pray for safety at the outset of a journey and then entrust one's life to God, or whether constant prayer showed a lack of faith. Discreet glances showed that my companions, like myself, were praying feverishly all the way back.

But the subcontinent had long been familiar to me from my father's Second World War memories. As a boy, he got on better with wildlife than with authority figures after a gamekeeper caught him trespassing with a butterfly net. Stationed along the Indus, he continued his winning ways with military leaders, who reprimanded him for loitering during battle exercises and parading with a rusty bayonet. Constitutionally incapable of inciting the spirit of the bayonet, a less martial figure seldom graced the theatre of war.

But it was the land that he remembered best, experienced in that first youth that does not return, in the words of novelist John Buchan. He recalled vividly 'my Indian jungle' in the Western Ghats and the brilliant Indian birds—the hoopoe, the golden oriole, and the hornbill. He loved the call of the barbet, the unseen 'coppersmith' with its monotonous ringing notes like a metalworker, 'the town crier to every Indian garden', said Rudyard Kipling, 'and tells all the news to everybody who cares to listen.'[4] And he never forgot the distant view of the Himalayan peaks, so enigmatic and so high above the horizon that they surely could not be solid rock.

Eduard Suess discovers how mountain belts work

In 1857, Eduard Suess approached the Austrian Minister of Culture and Education.[5] Could he be appointed as a junior unsalaried professor at the University of Vienna without an academic degree? The authorities of the university, where geology teaching was a dull inventory of specimens, had laughed at the question.

The teenaged Suess had taken part in the 1848 uprisings that swept Europe, fleeing to Prague after Austrian forces crushed the revolution. There, he was fascinated by fossils in the museum and surrounding countryside, but he was summoned back to Vienna and imprisoned for a month. Unable to return to the university, Suess organized fossil collections in what

later became the Natural History Museum, impressing Charles Lyell and other European experts with the depth of his knowledge.

The Minister, a learned man, glanced over glowing references from Austria's leading geologists. Regrettably, he said, the law was clear and Suess could not become a junior professor without a degree. However, continued the Minister, nothing in the law prevented Suess from becoming a senior professor. Appointed over the heads of his scandalized new colleagues, Eduard Suess became a professor at the age of 26, still drawing his museum stipend. In the words of biographer Celâl Şengör, the Minister single-handedly made a present of Suess to the scientific world.

Suess invented many geological terms that are so familiar that few know who invented them. They include *Gondwana-Land* and the *Tethys Ocean*; *eustasy* for the rise and fall of sea level; a *syntaxis* where a mountain belt curves round a continent; *shields* for the remnants of ancient cratons; and a universal concept—the *biosphere*. He was an outstanding teacher and gave generous encouragement to the young Jack Gregory to study the East African Rift Valley. Suess was elected to parliament but, a man of principle, resigned for a time from the city council in opposition to a public lottery ('he is a dear man', said a frustrated councillor, 'but a Professor!').

Above all, Suess was fascinated by mountain belts. His four-volume book *Das Antlitz der Erde* (*The Face of the Earth*)[6] took the geological world by storm with its global reach and stately prose. Science, like the world, wrote Marcel Bertrand in the French edition, was not created in one day, but when the history of geology is written, *Das Antlitz der Erde* will mark the end of the first day 'when there was light'. Mountain belts were usually considered symmetrical, long narrow tracts that subsided and rose where they stood. Charles Darwin had inferred that subterranean magma was forcing up the Andes. But Suess—a prodigious reader and correspondent—used the geometry of faults and folds to show that mountain belts everywhere were asymmetric, and horizontal compression had thrust enormous rock masses over each other. He focused especially on Asia, where the Himalaya was being thrust to the south.

How could this compression be explained? Suess drew on a model of a cooling Earth, popular at the time, in which the interior shrank faster than the rigid, wrinkling surface. *Das Antlitz der Erde* inspired Wegener, who admired Suess' global view and relentless consistency.[7] Suess died a year after Wegener's manuscript was published, and it is not known whether he learned of continental drift—a robust explanation for his observations.

And what of rivers? *Das Antlitz der Erde* implies that mountain belts evolve more rapidly than anyone could have imagined—an inference abundantly confirmed by later age dating. Drawing on the insights of Suess, Asian researchers showed how the growing mountain ranges rotated, reversed, and destroyed their rivers. Virtually no original Asian rivers survived the collision of India with Asia, and many are less than 20 million years old in their present courses.

India moves north

India moved north with lightning speed.[8] By the Cretaceous, the continent had parted company with Gondwanaland, and India began to collide with Asia some 60 million years ago in the Paleocene as the Tethys ocean crust was drawn down under Asia. By 50 million years ago in the Eocene, the Tethys Ocean had been eliminated between India and Asia, and India became an *indentor*, penetrating 2000 km into the solid rock of Asia as the crust thickened to form the high Tibetan Plateau (Fig. 6.3). Thrust sheets kilometres thick were forced upward and southward, forming the present line of Himalayan peaks late in the Oligocene some 25 million years ago as uplift outpaced erosion by water and ice. Mountain rivers poured a flood of sediment into a foreland basin to the south of the ranges, where the thrusts depressed the Indian craton and allowed thick river strata to accumulate. The mountain ranges wrapped around the corners of India to form western and eastern *syntaxes*, and *escape tectonics* forced faulted blocks eastwards, providing conduits for river flow to the Pacific Ocean.

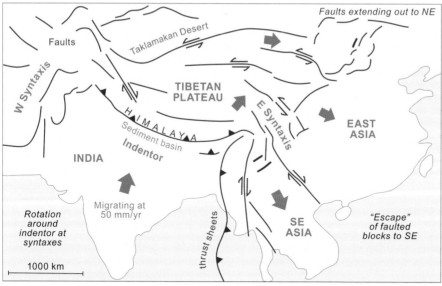

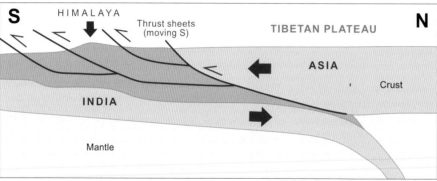

Figure 6.3 Tectonic map of southern Asia, showing thrust faults and other fault lines generated during collision between India and Asia. Cross-section shows southward-directed thrust sheets developed on the Indian Plate as it underthrust Asia. After Tapponnier et al., 2001.

Figure 6.4 Yamuna River at Kalpi, India, with 1948 monsoon level marked in red high up on the nearest railway bridge column. The marker is 18 m above the low-stage flow.

For millions of years the Southwest Indian Monsoon has been a powerful engine for the rivers, dramatically illustrated by a flood level at the Yamuna River (Fig. 6.4). Summer winds pick up moisture from the Arabian Sea, and intense precipitation tracks up the Bay of Bengal and onto the Tibetan Plateau, where rising warm air draws moist air up the deep river valleys towards a rain shadow north of the high peaks. There, the early morning air hangs cold and still, but suddenly springs to life as the wind roars up from the Indian plains, the dust whirling. Late in the afternoon the air pressure equalizes, and dusk descends in an eerie calm.

Mapping Asian rivers

Asia's empires were founded along the continent's rivers, part of what Indian writer Rabindranath Tagore called the 'strata of human geology'.[9] Connecting the cities through the mountain passes was the network of the Silk Roads, for some 6000 years the world's pre-eminent trading route, running from Xi'an in China to the Mediterranean.[10] The roads ran north of the Tibetan Plateau, skirting the southern fringe of the Gobi and Taklamakan deserts, and followed the Golden Road westward to Samarkand and south of the Caspian Sea. Connecting routes threaded the Himalayan river valleys and, west of the Indus, reached the Indus and Ganga plains through the Khyber and Bolan passes in Afghanistan with their dangerous flash floods. Islam spread from Arabia, Buddhism from northern India

and Nepal, and Nestorian Christianity from Syria. Along the Silk Roads swept the Mongol hordes of Genghis Khan and the Black Death, possibly carried by camels.

The Silk Roads saw walrus tusks transported from Siberia, lapis lazuli from Central Asia, and paper making, river lock gates, and iron-chain suspension bridges from China. The Jade Road brought to China the green stone that Confucius likened to the perfect human condition, strong as intelligence and moist and smooth as benevolence. And silk travelled west, discovered in Xi'an, it is said, when the wife of the Yellow Emperor unravelled a silkworm cocoon that had fallen into her tea. The Romans first saw silk in the banners of a formidable Parthian army in 53 BCE at the Battle of Carrhae in Mesopotamia, the last thing most of them ever saw. Silk was soon all the rage in Ancient Rome. Marco Polo followed the Silk Roads to China in the thirteenth century, and Moroccan traveller Ibn Battutah crossed the river passes to India and China in the following century, covering more than 100,000 km in his lifetime, probably the furthest by anyone before the steam age.[11]

Asian scholars had worked with maps and river courses for centuries, but as European nations sought colonies, accurate river maps became ever more vital policy instruments. In 1800, British administrators set up the Great Trigonometrical Survey which, over the next half century, formed the basis for mapping much of India and parts of the Himalaya. Carrying equipment on elephants under the command of martinet George Everest, the survey cost more and resulted in more deaths than the average small war as typhus, malaria, and tigers took their toll.[12] In 1856 a Himalayan peak was estimated at more than 29,000 feet and, with no known name at the time, was named for Everest.

Where did the Himalayan rivers rise? British administrators secretly sent native surveyors known as the Pundits over the mountains to find out.[13] The Pundits paced distances and kept records using rosaries, a standard item for Buddhist pilgrims on the Silk Roads. They hid their notes in prayer wheels with a concealed compass, and they carried sextants hidden in chests and thermometers concealed in staffs, returning with precise river locations. Pundit Kinthup secretly cut and marked five hundred logs to be put into the Yarlung Tsangpo for interception in the Brahmaputra if the rivers connected through the gorges of the Eastern Syntaxis. Unexpectedly enslaved for a time, he sent a message which was not received, and the logs likely floated to the Bay of Bengal. But by the turn of the century it was clear that the Yarlung Tsangpo ran into the Brahmaputra and not the Irrawaddy.

Rudyard Kipling made the secret surveyors famous in his 1901 novel *Kim*—a book that caused Mark Twain to declare that the Indian Subcontinent was the only foreign land that he daydreamed about. The young Kim was trained to

> make pictures of roads and mountains and rivers—to carry these pictures in thine eye till a suitable time comes to set them on paper … But, as it was occasionally inexpedient to carry about measuring-chains, a boy would do well to know the precise length of his own foot-pace.[14]

Russian and British interests in Asia clashed increasingly with those of China and Persia in what became known as the Great Game. Russian fortresses lined the Amu Darya, a natural invasion route to India, and alarmed British administrators prepared for invasion. In 1831 discreet soundings showed that the Indus was navigable for some thousand kilometres inland, and British and Russian spies and mappers engaged in a cold but at times cordial war across the Himalaya. By the early twentieth century, the courses of Asian rivers had been mapped in detail. But how and when they had attained their courses was a different matter.

River histories of the Indian Subcontinent

In 1919, Edwin Pascoe of the Geological Survey of India named the ancient *Indobrahm River* for a combination of the Indus and Brahmaputra.[15] For much of the time since Tethys closed, he asserted, the Indobrahm flowed westward along the line of the Ganga to the Arabian Sea, in the opposite direction to the modern river. Rivers draining to the Bay of Bengal then eroded headward and turned the river eastward to become the Ganga. What was left of the Indobrahm became the Indus and the rivers of the Punjab.

Shortly after arriving in India, Pascoe was sent up the Irrawaddy River to study the Burmese earth-

oil industry. This had been active for over a thousand years, extracting crude oil that oozed from rock fissures and could be extracted from hand-dug wells more than 100 m deep.[16] Over the next decade, he studied the Siwalik Hills in the Himalayan foothills in the hope of finding oil, and he managed the Indian Empire's oil resources. The Siwalik Hills were famous for their elephant, rhino, and early hominid fossils, preserved in strata of the Siwalik Group (Fig. 6.5). The strata were laid down by braided rivers on megafans that had emerged from the mountain front as much as 12 million years before in the Miocene.[17]

Geologists before Pascoe had puzzled over Asia's river history, drawing on concepts established in the 1870s by John Wesley Powell and his colleagues in western North America (see Chapter 9). In 1907, Richard Oldham speculated that India's drainage originally flowed north to the Tethys as *consequent* rivers, following the slope of the continent (Fig. 6.6) but reversing their flow to the south as the Himalaya rose.[18] The Yarlung Tsangpo and Indus were *subsequent* rivers that followed folds and faults in the collision zone, escaping round the syntaxes on either side of India. But *antecedent* rivers might be present, older than the mountains and cutting down to keep pace with the rising land. Other rivers might be *superposed*, established on a higher land surface but let down onto the present rocks as erosion proceeded. Mountaineer Lawrence Wager supported a superposed origin for rivers near Qomolangma,[19] but Oldham thought that the rivers had eroded through the mountains from the south, aided by a steep slope and intense monsoonal flow. There was much speculation and little certainty.

The ancestral Indus is now known to be among the oldest rivers of Asia, possibly draining to the closing Tethys and later flowing west along the line of collision between India and Asia (Fig. 6.7).[20] Seismic and well information from the offshore Indus Fan show that an ancestral Indus poured sediment into the Arabian Sea during the Eocene, more than 40 million years ago. The Yarlung Tsangpo may also be of similar antiquity, following the collision zone eastward. As India drove north into Asia, the Aravalli Range of Peninsular India began to impinge on the Himalayan front, dividing the Ganga from Pascoe's Indobrahm in the Miocene some 15 million years ago.[21] On its eastward route to the Bay of Bengal, the axial river picked up a series of steep,

Figure 6.5 Former sandbed rivers (pale strata) and floodplain mudstones (red) of the Siwalik Group, laid down on Himalayan megafans and later thrust southward (to the right) in the Siwalik Hills of India. The Siwalik Hills form the Himalayan Foothills north of the modern river plains.

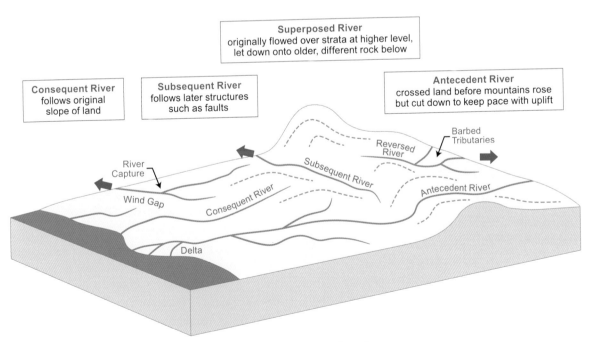

Figure 6.6 Inferred river histories, based on concepts developed on the Colorado Plateau by John Wesley Powell and colleagues in the late 1800s.

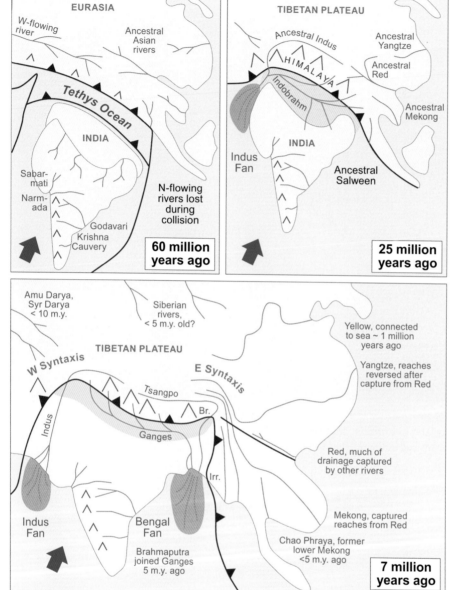

Figure 6.7 The evolution of some Asian rivers as a result of ongoing collision between India and Asia during the Cenozoic.

sediment-charged transverse rivers—the Gandak and Kosi, among others—that built impressive megafans from their mountain exits (Fig. 6.8). The Kali Gandaki in the Gandak headwaters is probably an antecedent river, maintaining its former course in recent times as it cut through the rising Himalaya.

The aggressive Brahmaputra, draining the eastern Himalaya, eroded headward and captured the Yarlung Tsangpo perhaps as much as 18 million years ago in the Miocene, cutting the latter's connection to the Irrawaddy (as Pascoe had suggested) and swinging the

river sharply through the Eastern Syntaxis to the Bay of Bengal (Fig. 6.9). Thereafter, the Yarlung Tsangpo and Irrawaddy populations of chameleon fish—capable of remarkable colour changes—evolved separately. As the Shillong Plateau rose athwart the river's course, the Brahmaputra was diverted westward about five million years ago in the Pliocene to a confluence with the Ganga.[22] The Amu Darya and Syr Darya are young rivers, perhaps ten million years old, that took drainage out to Central Asia as the Western Syntaxis grew.[23]

Figure 6.8 Ganga River showing meandering channel pattern, with the Kosi Megafan, 150 km long, and other megafans extending south from Himalayan exits. The megafan of the Son River (lower left) extends north from Peninsular India. Black areas of the river tract are water. Landsat 5 false-colour image from Fall, 2019, using a combination of red, green and blue bands (7, 5 and 4). Courtesy of Gary Weissmann and USGS.

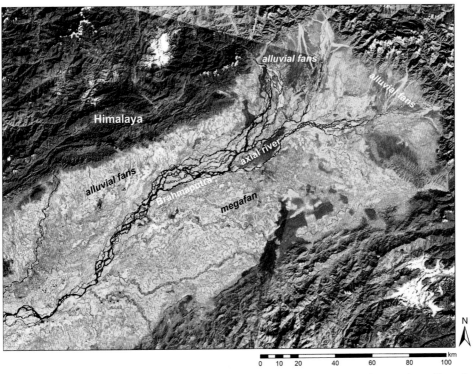

Figure 6.9 Brahmaputra River showing braided channel pattern, with large megafans and smaller alluvial fans and interfan tributary rivers extending from the Himalaya (northwest to northeast) and ranges in Myanmar (to southwest). Black areas in the river tract are water. Landsat 8 false-colour image from Spring, 2020, using a combination of red, green and blue bands (7, 6 and 5). Courtesy of Gary Weissmann and USGS.

In contrast to these relatively young rivers, the drainage systems of Peninsular India, not yet caught up in Himalayan earth movements, have broadly followed their present courses since Gondwanaland broke up.[24] Before the breakup, large rivers with headwaters in Antarctica flowed northwestwards across India, occupying Jurassic rifts as Gondwanaland began to fragment. As India drifted away from Antarctica and Australia in the Cretaceous and the enormous eruptions of the Deccan Traps domed up the Western Ghats, the Krishna, Godavari, and Cauvery rivers reversed their flow southeastward to the Bay of Bengal, still occupying the earlier rifts.

On the northwestern side of India, Cretaceous rifts opened as the continent broke away from Africa and Madagascar, guiding the Narmada and Sabarmati rivers to the Arabian Sea. As the Himalaya rose and the foreland basin subsided along the southern flank of the mountains, the northward tilt of the craton probably caused southward-flowing Narmada tributaries to reverse their flow to the Ganga, reducing the Narmada's catchment area. The Indian continent will continue to tilt and disappear below Asia for millions of years to come, possibly reversing the Sabarmati and Luni northward to the Ganga or Indus. But sooner or later, India will grind to a halt and Asia's rivers will achieve a transient peace.

River histories of eastern Asia

Jack Gregory decided to visit China.[25] Three big Asian rivers—the Salween, Mekong, and Yangtze—curve around the Eastern Syntaxis in deep gorges only a few tens of kilometres apart, after which they flow to deltas 3000 km apart in the Bay of Bengal, the South China Sea, and the Yellow Sea. A geographer once likened the pattern of converging and diverging rivers to a sheaf of thunderbolts in the hand of Jove. Running between them are the Chao Phraya and Red rivers, flowing out through Southeast Asia. In the early seventeenth century, Ming Dynasty traveller Xu Xiake linked an inland river to the Yangtze, extending the Yangtze's known course by 2000 km.[26] Tribal traditions thought that the upper Yangtze had once joined the Mekong.

Could the region's rock structure provide clues to the rivers' convergence? Gregory (the Chief) and his son Christopher (the Assistant) steamed up the Irrawaddy in 1922 and set out on a four-month excursion with a group of muleteers, porters, and an interpreter. It was a challenging expedition. The muleteers argued over payment, avoiding rocky passes and stopping in poppy fields to gather raw opium. The foraging mules crept away and were chased down by the muleteers with inhuman cries like yodelling monkeys. One mule fell off a bridge into the river, reducing their bread to a sodden mass. The region was rife with bandits, and district administrators provided escorts armed with rifles and umbrellas or spears and crossbows. Once, the Chief saw a head rise above the bushes and rushed ahead with a revolver to thwart an ambush.

The caravan passed through walled cities where mercury, silver, and gold were mined. They encountered mule caravans laden with salt and, for a while, travelled with a falconer and his hawk. Every valley had rice fields with canals and earth-dam reservoirs, bamboo aqueducts were common, and some towns had so many canals that Gregory considered them a Venice without gondolas. Some magistrates refused them permission to travel, and one official distrusted their maps until the interpreter explained that they were only cracking stones and putting marks on charts. They wisely forbade Gregory to carry boxes of silver to pay the muleteers and porters, courteously sending the money on main routes under military escort.

The scenery was magnificent. The Mekong was in full spate with booming eddies and crashing boulders like the roar of river dragons. In places landslides had torn away the roads, and some paths ran precariously on scaffolding attached to rock walls, with corners so sharp that the mules had to be unloaded to sidle round. The Chief fell into a river when the bank collapsed and the Assistant plummeted off a cliff path in the dark. Sliding across gorges on rope bridges was a special challenge, and after dark they groped along with an acetylene bicycle lamp or a lighted spar of pitchpine. They climbed over the walls of towns closed off to keep flood spirits out, and heavy rain diluted the gravy on their yak cutlets.

Exposed in the gorge walls were gigantic folds and thrusts, formed as Gregory put it 'when the wave of the Himalaya swept forward against the mass of Yunnan. Here on this old foreland the rocks could not but sing the epic of how, crushed and twisted, they had stood against the rocky breakers that foamed about them.'[27] Wind gaps and barbed tributaries testified to river capture and reversal. These were not antecedent rivers that had sawed through a rising mountain chain—they were subsequent rivers that followed the geological structures, their original courses long obliterated.

Gregory's findings broadly agreed with those of Bailey Willis, who had worked with John Wesley Powell and had travelled down the Yangtze in 1903. The Yangtze might originally have been superposed, he thought, but its course was now controlled by faults and folds, perhaps uniting once-separate rivers. The river was 'imprisoned by tremendous canyon walls, which its own flood has carved and is still carving. Whatever channel it once had … may be traced only above the great cliffs among the mountain summits.'[28]

Having won out over the grumbling muleteers, Gregory was carried in a sedan chair for the last days of the expedition. Reclining in a deckchair on an Irrawaddy steamer, he planned a popular account of the excursion, which had followed paths traversed by Marco Polo. Gregory was the first to use the term 'Silk Road' in English.

Later research began to document the river histories.[29] From the coastal ranges of China, rivers flowed west to the Tethys in Central Asia before India collided with Asia and the rise of the Tibetan Plateau forced a reversal of flow to the Pacific. Within as little as 15 million years after the collision, the Tibetan Plateau was high above sea level, still preserving in places relict drainage patterns that may be the ancestors of the modern rivers. Many of the relict patterns are dendritic, suggesting that the region was once a low-relief continental interior where rivers dissected relatively uniform strata. As the Tibetan Plateau continued to rise and the rock masses were forced to the east along faults, the rivers were brought closely together around the Eastern Syntaxis, cutting gorges several kilometres deep that locked the rivers in place. Increasingly distorted by the ongoing deformation, few rivers survived the upheaval in anything like their original courses.

The Red River may initially have been dominant but, as the monsoon intensified, the Yangtze, Mekong and Salween, which started life as small coastal rivers, eroded headwards and captured much of the Red system. The middle part of the Yangtze, once a tributary to the Red, reversed its flow for over a thousand kilometres until, sometime in the Miocene, the river ran through the famous Three Gorges and its course

became much as it is today. The lower and upper parts of the present Mekong became connected, and about five million years ago in the Pliocene, its former course through Thailand became the Chao Phraya River.

Out to the north the Siberian rivers may be only a few million years old in their present courses to the Arctic and Pacific oceans.[30] But their ancestors may have flowed north across Eurasia before Tethys was consumed.

Yellow River of China

The Yellow River is more than 5000 km long, running through the upland plains of the Tibetan Plateau, dotted with herds of yaks, before swinging unexpectedly around the Ordos Loop and across the North China Plain to the Yellow Sea. The Mongol emperor Khubilai Khan sent an expedition in 1281 to find the river's source.[31]

But it is not for its flow that the Yellow River is famous. Each year the river transports an extraordinary 120 million tons of suspended sediment to the sea. In some years each cubic metre of water contains 35 kg of sediment, compared with ten for the Colorado and one for the Nile. The Yellow owes this enormous load to the Loess Plateau within the Ordos Loop, the world's largest archive of wind-blown dust. Hundreds of metres thick, the soft yellow silt or *loess* covers an area larger than Germany.[32] As desert conditions set in across central Asia more than eight million years ago, intense winds swept dust from dry lakes and rivers out to the plateau and the Pacific. Streams and gullies erode the soft silt, and the unstable Yellow River has built up sediment at a frantic rate. Its bed now stands ten metres above the North China Plain, held in place by unstable dikes that have ruptured catastrophically (see Chapter 16). Since a major breakout in 1855, the channel has shifted nine times into new delta lobes.[33]

The Yellow traverses a geologically active area, and ancestral river reaches may have followed faults in the Ordos Loop since the Eocene.[34] Changing slopes have reversed the flow, and lakes have divided the river into disconnected segments. The river carved wide *peneplains* ('almost plains') across the landscape during stable periods but cut deep gorges with terrace staircases as faults moved and the region rose (Fig. 6.10A,B). Perhaps less than one million years ago, a long-lived lake on the edge of the plateau broke out through a

Figure 6.10A Loess Plateau of northern China. Agricultural terraces are developed on river terraces with gravel (tough projecting beds in small cliffs at left) overlain by fertile loess. The gravel was laid down by Yellow River tributaries during tectonic uplift and incision. Flat-lying surface on bedrock in the distance is a planation surface > 10 million years old, cut laterally by the Yellow River during a period of relative stability. The river then incised through the rock to form a deep valley, out of sight in middle distance.

Figure 6.10B Terrace of the Yellow River, with well-rounded river gravel resting erosionally on underlying strata. The terrace gravel formed during tectonic uplift as the river cut down and stranded its former sediments.

gorge to the Yellow Sea, finally bringing a unified Yellow River to the ocean. A connected Yellow River is one of Asia's youngest rivers.

Asian rivers of the soul

Hermann Hesse found Asia disappointing. His parents had lived in India, and the subcontinent was as much part of his interior landscape as his native Germany.[35] Arriving in Asia in 1911, he briefly shuffled off a disturbed childhood and a difficult marriage, but he was distracted by the chaos of noise and dirt in bustling towns surrounded by what he described as the wild, lustful jungle. Although he detested colonialism, he could not overcome his European bias and went home before reaching India. But his Asian travels provided Hesse with an imaginary landscape, and the mosaic of coexisting cultures had a spiritual unity that seemed lacking in Europe. Sessions of psychoanalysis created a peaceful inner space for the completion in 1922 of *Siddhartha*, which spoke strongly to young people in the aftermath of war.

In *Siddhartha*, Hesse used rivers as a powerful source of imagery in the conversations of Siddhartha and the ferryman Vasudeva. 'Yes', said Vasudeva, 'it is a very beautiful river. I love it above everything. I have often listened to it, gazed at it, and I have always learned something from it. One can learn much from a river.' In a moment of insight in his despair, Siddhartha understood 'that the water flowed and flowed and yet it was always there; it was always the same and yet every moment it was new.' 'Is this what you mean?' asked Vasudeva. 'That the river is everywhere at the same time, at the source and at the mouth, at the waterfall, at the ferry, at the current, in the ocean and in the mountains, everywhere, and that the present only exists for it, not the shadow of the past, nor the shadow of the future?'[36]

A year before Hesse's journey, Rabindranath Tagore, scion of a leading Bengal family, published *Gitanjali*, receiving the 1913 Nobel Prize for Literature.[37] Written by the Padma River after the deaths of his father, wife, daughter, and son over the previous few years, many of the poems explore Tagore's feelings through river imagery. Left in a London train and reclaimed from the Lost Property Office, the manuscript was given to W.B. Yeats, who read the poems to an admiring group of British writers. Poems from an Asian river resonated with people around the world.

Tagore knew the rivers intimately. His grandmother had lived in a hut by the Ganga and, on the night before her death, Tagore's father had a profound experience alone by the river. In 1891, Tagore became manager of the family's Padma estates, living for a time on a houseboat. 'I stood for a moment at my window,' he wrote, 'overlooking a market place on the bank of a dry river bed, welcoming the first flood of rain along its channel. Suddenly I became conscious of a stirring of soul within me … a sudden spiritual outburst from within me which is like the underground current of a perennial stream unexpectedly welling up on the surface.'[38] 'The heaven's river has drowned its banks and the flood of joy is abroad,'[39] he wrote in *Gitanjali*.

Hesse and Tagore knew their rivers, across landscapes and cultures. And, in my imagination, the Indian coppersmith that delighted my father still sounds its rhythmic notes from the treetops.

CHAPTER 7

THE CONFLICTED RIVERS OF EUROPE

Napoleon trades geology for geopolitics

Napoleon Bonaparte gazed across the Mediterranean Sea from his hilltop residence at Portoferraio on the island of Elba. To the east he could see the Italian mainland and far to the north lay France, from where he had recently been exiled. His eye was caught by a white cliff shining in the sunlight with a beach of pale pebbles at its foot (Fig. 7.1A). The Little Emperor smiled to himself and turned away.

Within a few days of his arrival in 1814, the energetic Napoleon had toured the island with its ancient iron mines and, seemingly resigned to his fate, had set out to govern his new kingdom. He would give his life, Napoleon declared, to science and literature. 'I am now a deceased person,' he noted, 'occupied with nothing but my family, my retreat, my house, my cows, and my poultry.'[1]

Had Napoleon devoted his formidable skills to science and the inhabitants of Elba, much might have been accomplished. But denied a promised French annuity, he became restless. An apparently ludicrous palace protocol served as a protective screen to outwit surveillance and, eventually, a patrolling British frigate. Landing on the French coast, Napoleon ousted the Bourbon nobility who had returned to power with their accustomed *hauteur*, before going down to defeat at the Battle of Waterloo. Thereafter he exchanged the white cliffs of Elba for the dark basalts of St. Helena in the South Atlantic.

The white cliff and beach pebbles near Portoferraio consist of *eurite*, a microgranite studded with blue-black crystals of the mineral tourmaline (Fig. 7.1B). Writers in antiquity identified the black spots on the pebbles as drops of sweat from the Greek Argonauts after they rowed to the island or conducted athletic games. The rock crystallized from rising magma nearly eight million years ago, one of many responses to the jostling of Africa, Europe, and smaller plates that generated the Alps, Apennines, and other mountain ranges.[2] To the south,

Figure 7.1A Portoferraio, Isle of Elba, with Forte Falcone in foreground and Napoleon's residence in exile on the headland at left. Beach at right has white pebbles of eurite.

Figure 7.1B White pebbles of eurite, a microgranite with dark spots of tourmaline, on the beaches at Portoferraio.

another intrusion forms the island of Monte Cristo, the treasure island in the famous book by Alexandre Dumas, centred around Napeoleon's return from Elba.

Between six and seven million years ago, a large igneous intrusion in western Elba domed up an eroding landscape that included the eurite intrusion, and

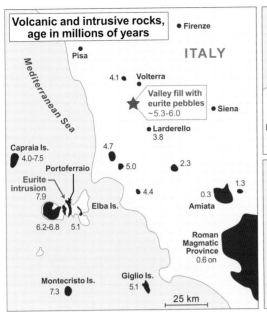

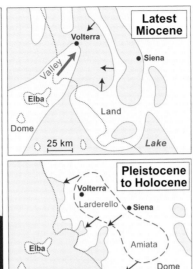

Figure 7.2 The age of volcanic activity in mainland and offshore parts of Italy. Igneous doming of Elba caused eastward drainage, but this was reversed following doming of Larderello and Amiata. After Pascucci et al., 2006.

rivers carried the distinctive eurite pebbles eastward to the Italian mainland (Fig. 7.2). In cliffs within sight of Volterra, the pebbles are preserved in a Miocene conglomerate that fills a former valley 60 m deep and two kilometres wide, cut through lake clays. The river valley was swept by impressive floods that built up thick accumulations of boulders, and the eurite pebbles are a 'smoking gun' that links the river to Elba. To collect samples, my enterprising colleague Vincenzo Pascucci abseiled down the conglomerate cliff on a rope tied around a tree. An expert operator, it is rumoured that he can get a hotel built to help a conference—also, perhaps, a road and a bridge. Later, volcanic activity moved eastward to the Apennine mountain chain of the Italian mainland, where intrusive centres at Larderello and Amiata domed up the landscape. As the land rose, the rivers reversed their flow to run westward to the Mediterranean as they do today—a testimony to the rapid response of rivers across Europe to subtle changes in elevation and slope. Europe's landscapes are dynamic and many of the rivers are young.

Danish anatomist Nicolaus Steno famously studied the geology of Volterra and Elba while working for the Medici scientific academy in Florence.[3] The young Steno had publicly dissected the head of a great white shark from the Mediterranean, realizing that the formidable teeth resembled the miraculous 'tongue stones' in the rocks of Malta. Steno found that the town walls of Volterra were rich in fossil shells. He drew on his knowledge of the region in his 1669 publication *Dissertationis Prodromus*, describing the rock strata as an archive of deep time, in which he observed horizontal layers laid one upon another and later tilted. Unconcerned about empire, Steno became an advocate for the poor, and Pope John Paul II beatified him in 1988 for his scientific contributions and saintly life.

Competitive Alpine rivers: Danube, Rhone and Rhine

In the view of Swiss geologist Rudolf Trümpy, Plate Tectonics should have been invented in the Alps, the heartland of Eduard Suess and Alfred Wegener, where large folds and thrust faults were first described. He suggested that the modest navies of Switzerland and Austria (landlocked nations) had something to do with it because the theory, like Venus, was born out of the sea and was beautiful.[4] The young mountain chains of Europe came into being as a result of plate collisions after Pangea broke up. By the Cretaceous a group of small continents backed by Africa were starting to collide with Europe as Tethys closed, forcing up the Alps about 30 million years ago.[5] To the east, the Carpathians emerged from Tethys some 15 million years ago, ballooning out far beyond the Alps, and the Iberian plate collided with Europe to form the Pyrenees.

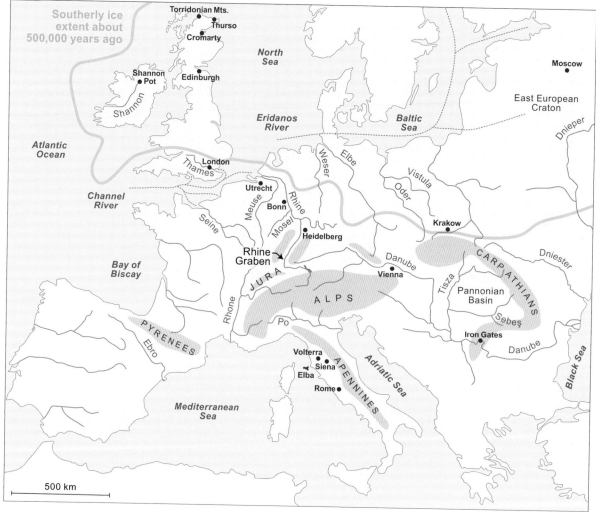

Figure 7.3 Modern rivers of Europe. Southerly ice extent from Cordier et al., 2017.

A volcanic rift system ran from the Mediterranean up the Rhone and Rhine valleys and into the North Sea. Tethys was reduced to brackish gulfs, leaving only the Mediterranean, Black Sea, and Caspian Sea to mark its former extent.

As in Asia, the rivers responded magnificently.[6] Alpine drainage is largely a tale of three competitive rivers, the Rhine, Rhone, and Danube, which rise close together but flow to different seas (Fig. 7.3). The Rhine flows into the Rhine Rift Valley and out to the North Sea, the Rhone flows around the western Alps to the Mediterranean, and the Danube flows around the eastern Alps and inside the Carpathian loop to the Black Sea. All three are subsequent rivers that follow the deforming rock structures and are less than ten million years old in their present courses.

As the Alps rose, sediment poured from a growing upland that was initially an island within Tethys, and the rivers that once flowed south to Tethys from the Vosges, Black Forest, and across northern Europe gradually became tributaries of Alpine rivers (Fig. 7.4). By ten million years ago in the Miocene as the Jura mountains rose, the ancestral Danube was flowing east along a subsiding foreland basin at the mountain front (Fig. 7.5). The river filled the Vienna Basin with Alpine sediment and then forged a route through Budapest (Fig. 7.6), building a delta that advanced for 400 km in six million years across a deep lake in the Pannonian Basin. The river cut gorges with terrace staircases, captured other streams, and exploited paths through the mountains, eventually flowing out through the Iron Gates gorge and connecting Alpine

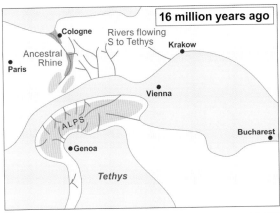

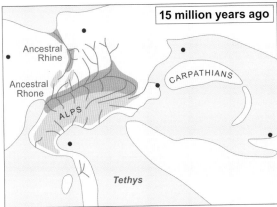

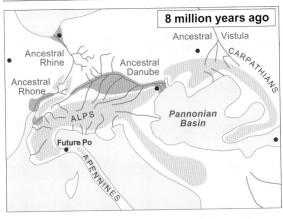

Figure 7.4 Evolution of rivers in the Alpine region, linked to collision between Europe, Africa, and several small plates. As the mountain belts rose, rivers flowing south to Tethys were incorporated into the developing Danube, Rhine and Rhone systems, and the Vistula flowed north from the rising Carpathians. After Kuehlmann, 2007.

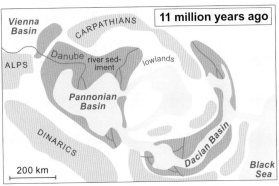

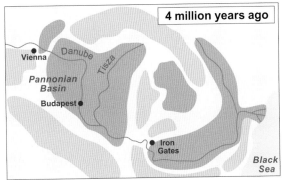

Figure 7.5 Stages in the evolution of the Danube River. With a large load of Alpine sediment, the river delta advanced into a palaeolake in the Pannonian Basin, and cutting gorges through the Iron Gates and elsewhere, connected to the Black Sea about four million years ago. After Olariu et al., 2018.

Figure 7.6 Danube River in Budapest, Hungary.

drainage to the Black Sea about four million years ago in the Pliocene.

The Rhone is also relatively old and may initially have been the pre-eminent Alpine river. For at least the past ten million years, an ancestral Rhone has flowed around the western Alps, periodically shifted by earth movements. When the Mediterranean was reduced to a shallow brine sea, the Rhone carved a valley two kilometres deep in as

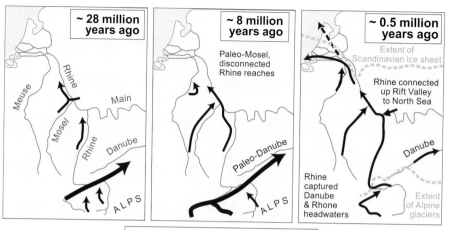

Figure 7.7 Stages in the evolution of the Rhine River. Comprising disconnected drainages for much of its history, a connected Rhine from the Alps to the North Sea may be as little as three million years old. After Preusser, 2008.

Figure 7.8 Mosel River, a tributary of the Rhine and one of the earliest rivers in what later became the Rhine system.

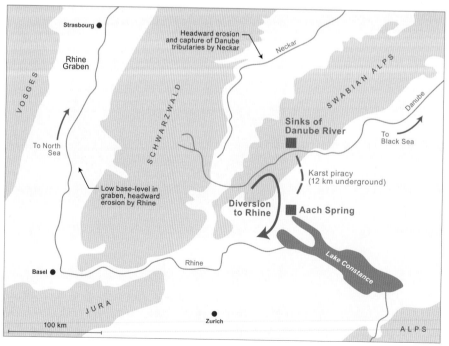

Figure 7.9 Progressive capture of the Danube headwaters by the Rhine and its tributaries, including underground drainage to the Rhine through a limestone area. After Hötzl, 1996.

little as half a million years, like the Nile filling the valley with sediment as the Mediterranean rose again.

A connected Rhine from the Alps to the North Sea is the youngest of the three rivers, frequently disrupted by rift faults and volcanoes (Fig. 7.7). River sediments in the area date back to the Eocene some 35 million years ago but were laid down by small disconnected rivers, including an ancestral Mosel flowing in from the west (Fig. 7.8). A river in the lower Rhine area carved a connection headwards through the uplands of the middle Rhine, and the Rhine finally captured its Alpine headwaters from the Danube and Rhone as recently as three million years ago, bringing Alpine sediment to the North Sea for the first time. With its low elevation in the subsiding rift, the Rhine will continue in an act of piracy to steal flow from the Danube, part of which trickles into the Rhine through caves and springs, leaving the upper Danube dry at times (Fig. 7.9).[7]

Other European rivers are also relatively young. By eight million years ago, the Vistula River was flowing north across Poland from the rising Carpathians. Less than three million years ago, the delta of the Po River of Italy advanced 350 km eastward in a million years as glacial floodwater from the Alps and Apennines swept sediment down its tributaries. South of the Pyrenees from the Cretaceous to the Eocene, an ancestral Ebro River flowed northwest to the Atlantic until Pyrenean megafans built out westwards in the Oligocene and Miocene and earth movements blocked the Atlantic exit. For a while, Ebro Basin drainage was endorheic and terminated in lakes, until a steep Mediterranean coastal river captured the Ebro some 14 million years ago in the Miocene, aided by rising lake waters that spilled over the divide.

And out to the north of the mountain chains, the European crust rose slowly through the Cenozoic at about one-tenth of a millimetre a year. Gentle folding created the arch of the Weald–Artois Anticline, which was later cut through to form the English Channel (see Chapter 14). The Meuse River, which had existed since the Eocene, cut down through the rising Ardennes, affected by rifts and volcanic doming in the Eifel area. As the land rose, antecedent rivers that included the Meuse, Seine, Somme, and Thames cut down to leave staircases of terraces, where Stone Age peoples left hand axes among the river gravels.

Winter Reise along the Rhine

Dressed in a black gown, white collar, and academic cap, I process with my Dutch fellow judges behind an official with a mace into the examination hall at Utrecht University. We formally address the doctoral candidate, Sanneke van Asselen, as Dear Candidate, and each judge asks nine minutes of questions. Questions about river channels in peat. Questions about how rivers built the Rhine–Meuse delta. Dear Candidate answers with poise. After precisely 45 minutes, the official knocks

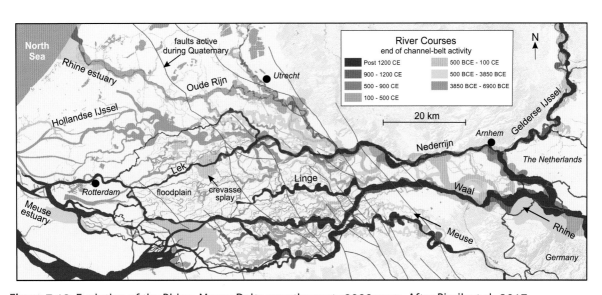

Figure 7.10 Evolution of the Rhine–Meuse Delta over the past ~9000 years. After Pierik et al., 2017.

her staff on the floor, declaring *'Hora est'* ('It is time'). We process out and agree on a rarely given *cum laude*. Dear Candidate receives a scroll and a eulogy is read. Freed from academic robes, we drive to Amsterdam for a party in which fellow students act out Dear Candidate and her research committee with thoughtful caricature.

Yesterday, though, we were out on the Rhine-Meuse delta, a flat landscape of windmills and distant church towers, much of it below sea level. Over a forty-year period, Utrecht scientists Henk Berendsen and Esther Stouthamer and their students drilled 200,000 sediment cores as they mapped river channels like tangled sphagetti below the delta surface (Fig. 7.10).[8] The channels date back nearly 9000 years, and the rivers avulsed frequently into new courses on the low-gradient delta plain, often where active faults intersected the drainage. Protective dikes often fail where they cross buried channel sands, and the maps aid in assessing hazards from floods that can scour holes 20 m deep through the dikes. Dear Candidate was plastered with mud as she raised a core of fibrous dark peat, the delta plants of earlier millennia, which is resistant enough to lock the channels in place.

It is now January and I am making a *winter reise* (winter journey) by train down the Rhine. The poems of *Die Winterreise* were written by Wilhelm Müller, who survived four Napoleonic War battles, and they record the journey of a lover, paralysed by grief, across a frozen winter landscape: 'My heart, do you now recognize / Your image in this brook? / Is there not beneath its crust / Likewise a seething torrent?'[9] A few years later Müller died suddenly, perhaps from a prolonged illness, after sailing down the Rhine past Lorelei Rock, where the legendary sirens drew boatmen to their doom. The poems resonated with the terminally ill Franz Schubert, who set them to music just before he died.[10] It was Müller's son Max who translated the *Rigveda*, among the world's earliest religious writings, and introduced Europe to the lost Saraswati River of the Indian subcontinent (see Chapter 15).

The university town of Heidelberg where Alfred Wegener practised fencing is covered by a drift of snow that gives definition to the ruined castle and the cobbled streets. Barges pass along the Neckar where the bridge records dramatic former flood levels, and in a darkened church, an organ scholar sits in a pool of light, filling the church with music. Seen from the train, the upper Rhine below Heidelberg is a manufactured but beloved landscape, a country of lines—vine rows marching up the hillsides, the furrows of ploughland, a line of crows on a hedgerow. The river enters the deep gorge of its middle reach where vineyards cling to the rock walls and where I am the only visitor to fairy-tale, snow-shrouded castles. The train whistles past Lorelei Rock, and a colleague takes me to find Devonian plants in a quarry deep in snow by the Mosel River.

The Rhine flows on through the industrial heartland of Germany to the North Sea. I visit Cologne where the millennial European flood of 1342 carved gullies 10 m deep in the fields—an agricultural disaster that may have contributed to the later devastation of the plague.[11] An even higher flood later in the century topped the city walls as boats drifted into the town. At Bonn I peer into the room where Beethoven was born, and a busker plays Bach on an accordion under a bridge. It has been a cheerful and splendidly musical *winter reise*.

Rivers of Russia and Arabia

Sir Roderick Murchison was an energetic Scot.[12] At the age of 16 he had fought in the Napoleonic wars in Spain and Portugal, raising the regimental standard from a fallen comrade under the eye of the Duke of Wellington. During the retreat to A Coruña, the exhausted Murchison was almost captured by Napoleon's troops but, aided by a swig of wine, he gathered the strength to get away. He enlisted as a cavalry officer, but Napoleon was defeated before he could get into the action.

A wealthy son of the establishment, Murchison soon settled down to hard riding and fox hunting. While shooting partridges with the chemist Sir Humphry Davy, Murchison realized that he could maintain his outdoor interests while pursuing a career in science. He visited Oxford professor William Buckland whose house was littered with rocks and bones, and Buckland's interpretation of the Thames Valley appealed to Murchison's feel for landscape from his military and hunting days. On a field excursion in France, he walked the legs off the younger Charles Lyell.

As a colleague noted, Murchison reforged his sword as a geological hammer.[13] By the end of his formidable career, he had held the offices of Director-General of the Geological Survey and President of the Royal Geographical Society—a general of science, as Napoleon and

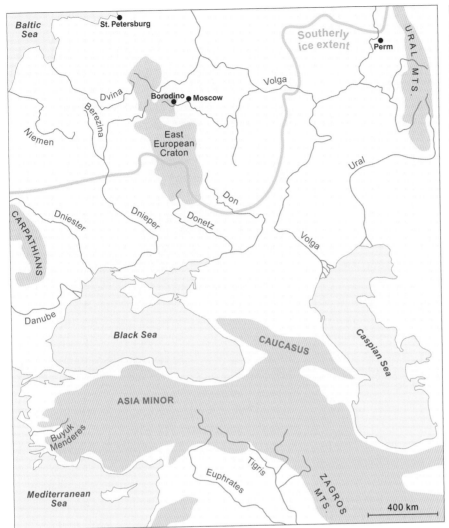

Figure 7.11 Modern rivers of Eastern Europe. Rivers from the East European Craton and Ural Mountains flow south to the Mediterranean, Black, and Caspian seas, which are remnants of the Tethys Ocean. Southerly ice extent from Astakhov et al., 2016.

Wellington were of armies. Having defined the Silurian and Devonian systems of strata in Britain, Murchison decided to track the strata into Russia, which appealed to him as an imperial society in which all classes knew their place. An experienced campaigner, he carried in his wagon a folding iron bedstead, a scarlet dress uniform, a sword, and Madeira wine. Accompanying him in 1840 and 1841 were two fellow scientists, Cossack guards, and a valet called 'Ivan the venerable'.

En route Murchison stopped in Berlin to visit Alexander von Humboldt, who in 1829 had travelled across Russia to the borders of China.[14] Treated by the Czar as a personal guest, Humboldt had been tactfully silent about his liberal political views but used an audience with the Czar to plead for persecuted groups. Murchison's expedition depended on his friendship with

Czar Nicholas I, whom he considered an enlightened ruler and who was seeking new mineral resources and, with railroads in mind, coal. Thoroughly in his element, Murchison commented frankly in his diary about the magnificent bosoms of court ladies and the gentlemen's tight legwear that displayed the 'virile member' to good effect. He watched the Czar review a body of 45,000 troops and attended palace balls, revelling in banquets with thirty varieties of wine and five classes of waiter.

The geologists moved south along the Volga River (Fig. 7.11), galloping to cliffs of red strata or sailing past them accompanied by forty musicians with champagne and musket volleys. Landowners treated them to feasts of caviar washed down with vodka, and he celebrated the Emperor's birthday in his cavalry uniform, receiving garrison reports as a presumed English general. Passing harrowing processions of manacled prisoners trudging towards Siberia, Murchison was thankful that England could banish convicts by sea.

In the gorges of the Ural Mountains, the strata were 'thrown up in vertical beds to form peaks, there coiled over, even like ropes in a storm, or fractured and broken in every direction'.[15] He found the lack of maps a handicap: were he the Czar, 'at least one thousand of my lazy officers would work for their laced coats, and produce me a good map, or they should study physical geography in eastern Siberia!' By imperial order, a reservoir to assist iron smelting was drained to allow him to canoe down a river, where he capsized in a furious flood. On the mountain summit he stood with one leg in Europe and one in Asia and sang 'Long live the Emperor!'

On the Russian Steppe, the carriages sank to the axles in black mud at river crossings, and Murchison swore like a Russian general to get them pulled out. On night journeys he slept rolled in his cloak, surviving on potted lobster and tobacco. Peasants and soldiers alike required amazingly little to make them happy, and he began to describe the steppe peoples more and judge less by his accustomed standards.

After a parched summer, dead horses lay along the road back to Moscow and St. Petersburg like the debris of a retreating army. After waiting in an anteroom with a painting of Napoleon retreating across the Berezina River, Murchison was presented with a state decoration by the Czar. The expedition had been remarkably successful, and Murchison and his colleagues published *The Geology of Russia* in 1845,[16] defining the Permian (named for the medieval kingdom of Permia) for the red rocks along the Volga River, as well as mapping younger strata that set the stage for understanding the region's river history.

The strata that Murchison mapped lie on the bedrock of the East European Craton, more than three billion years old, which collided with the Siberian Craton to form the Ural Mountains in the Carboniferous and Permian. Incredibly, some of the world's oldest rivers still run broadly in their ancient courses across this region. In the Jurassic, the ancestral Volga, Don, and Ural rivers flowed south, along with the Donetz and Dnieper—consequent rivers that followed the original slope of the land southward from the Urals and the East European Craton to the Tethys Ocean.[17] Some

river reaches may occupy courses established in the Carboniferous more than 300 million years ago. These ancient rivers still flow to the Black and Caspian seas, remnants of Tethys. Here, too, the Messinian lowering of sea level probably played a part in the river histories, causing the Volga to carve out a gorge 600 m deep that ran for 2000 km inland, now filled with sediment.

The collision zone between Eurasia and Arabia is represented by the active tectonic belts of Asia Minor, the Caucasus, and the Zagros Mountains, the latter rising as a southward-directed thrust zone in the Eocene about 35 million years ago and generating the subsiding foreland basin of Mesopotamia. As the mountains rose, the Tigris and Euphrates flowed south from the collision zone, drawn into the foreland basin.[18] The Euphrates was close to its present course nine million years ago in northern Syria, where its gravels are capped by a dated lava flow. As Arabia rose during the Miocene and Pliocene, the river advanced 800 km across the bed of a large lake to the Persian Gulf, cutting through the bedrock in a deep gorge and forming a spectacular terrace staircase near Raqqa.

And out in northern Norway, the little Tana River flows into the Barents Sea across land scoured by the ice sheets. A young river, one might suppose, occupying a new course after the Ice Age. But drilling below the Barents Sea has identified a thick accumulation of Triassic sandstone and mudstone laid down by a large river that ran north from the Ural Mountains in their heyday. An ancestral Tana flowed across the Baltic Shield, an extension of the East European Craton, to join this large river.[19] Despite interruptions from Mesozoic sea-level rise and glaciation, the little Tana still follows a course that came into being some 250 million years ago, making it among the older rivers on Earth.

Napoleon crosses the Berezina

Napoleon stood on the bank of the Berezina River, wrapped in his overcoat.[20] How had it gone so wrong? Only five months before, his Grande Armée of half a million troops had crossed the Niemen River, opposed only by a Cossack who came down to ask what they were doing. The army had advanced towards Moscow along a corridor between the Dvina and Dnieper rivers, and his troops had fought the Russian army to a stalemate at the Battle of Borodino, at a combined cost of 75,000 casualties. When Napoleon reached Moscow expecting a surrender, he had found the city deserted. Soon after it went up in flames.

And then the strategic withdrawal—not, of course, a retreat. Marching back past the now-pointless Borodino battlefield with its putrefying corpses, the soldiers were harried by Cossack horsemen and broken by winter conditions. They reached the Berezina, a tributary of the Dnieper, only to find the river a treacherous meltwater flood and Russian forces at the bridge crossing. They were trapped.

Then a French officer saw a mounted villager ford the river north of the bridge. The officer took his horsemen over, but the river would have to be bridged for infantry and artillery.

Napoleon's chief engineer, General Eblé, deployed his 400 men, mostly Dutch sappers, to build two trestle bridges using timbers torn from nearby houses. Working under fire and up to their necks in the icy water,

few survived. Some 30,000 troops crossed over until the bridges were set ablaze as a mob of stragglers rushed to safety. The crossing cost Napoleon some 25,000 casualties and as many more among the stragglers.

But the Berezina crossing had been one of his greatest achievements. It had all been the fault of the Russian winter, and now it was his duty to return to Paris. After all, he declared later, 'A man such as I am does not concern himself much about the lives of a million men.'[21]

It was left to Marshal Ney to march the shattered troops back to the Niemen. He got 2000 men over the Dnieper by jumping across the ice floes, beating off Cossack attacks through snowy ravines. At the Niemen, as few as 5000 men staggered across the bridge while Ney mounted a rearguard action. At night with the Cossacks approaching, he burned the bridge behind him, the last man of the Grande Armée to cross out of Russia.

CHAPTER 8

THE REVERSING RIVERS OF SOUTH AMERICA

The Andes and the River Sea

In 1541, Francisco de Orellana sailed down a river on the eastern flank of the Andes on a hastily constructed ship, searching for food to save the starving Spanish expedition of which he was a member. Reaching a confluence with the Amazon, he realized that he would have to march back overland to rejoin his comrades.[1]

But fame was knocking on his door. Encouraged by an eager crew, he sailed on down the Amazon. Six months later they became the first Europeans to follow the river to the Atlantic Ocean, eating skin belts and leather shoes boiled with herbs to survive. Along the way, de Orellana encountered large indigenous populations, and he named the river for female warriors whom he claimed to have met and who resembled the Amazons described by Herodotus. Reaching Spain with tales of El Dorado gold, he returned to the Amazon but died in the rainforest in 1546.

Twenty years later, Lope de Aguirre sailed through the Amazon system. He also passed many indigenous villages but, a brutal psychopath, his interests lay elsewhere. As a chronicler noted, 'he was growing morose because many days had elapsed since an occasion had offered to kill any one'.[2]

It was a priest, Pedro Cristóbal de Acuña, who provided the first serious account of the Amazon and its inhabitants. He described a line of villages along the river bluffs, so closely spaced that workmen in one village could be heard in the next. A town at one river mouth was reported to support 60,000 archers.[3] However, European oppression and disease soon decimated the inhabitants, taking much of their unique knowledge of plants and river landscapes with them.[4]

The Amazon is *the* river—*paraná*, a vast inland sea that carries a fifth of all the river flow on Earth.[5] With a catchment of more than six million square kilometres and a depth of 100 m in places, the river has an average flow rate at its mouth equal to ninety Rhines or eleven Mississippis and more than the next eight largest rivers combined (Fig. 8.1A). So vast is the normal flow that mariners far offshore find themselves sailing through a freshwater ocean, where river fish can migrate along the coast for hundreds of kilometres to enter other rivers. The Amazon boasts the largest recorded rainfall flood of 370,000 cubic metres a second—sufficient to supply the United States' entire daily water needs for irrigation, hydropower, and municipal use in one hour. At high tide the river-level rises for 800 km upstream to Santarém, turning the downstream parts of tributaries such as the Tapajós into lakes (Fig. 8.1B).

The Amazon carries more than a billion tons of suspended sediment to the ocean each year, far more than any other river. The mountains yield *white water* loaded with sediment and nutrients, while the lowlands yield *black water* darkened like weak tea by organic chemicals leached from the rainforest litter. Where whitewater and blackwater channels meet, varicoloured turbulent plumes roll down the river.

Why do long mountain ranges, named *Cordillera* from the Spanish word for rope, run down the western side of North and South America?[6] The ranges were not generated from continental collision. However, Wegener reasoned that if the Atlantic is opening and the Americas are moving west, mountains should rise where their western edges encounter a resistant ocean crust. It was a thoughtful insight: later research has shown that the history of the South American Cordillera broadly matches that of the opening Atlantic. During the past 150 million years, an extraordinary 13,000 km of Pacific crust has been subducted below the advancing Americas. A flotsam and jetsam of continental fragments and volcanic islands, like those that litter the Pacific today, are hung up along the edge of North and South America.

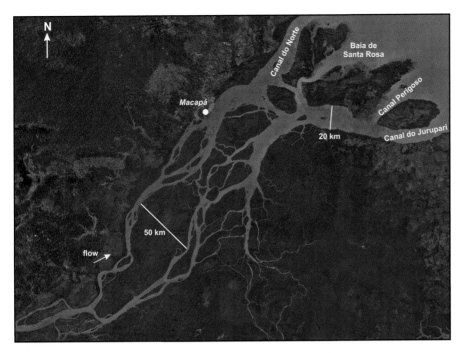

Figure 8.1A River Amazon at the Atlantic Ocean, showing distributaries and large plume of sediment-charged water. Maps data: Google, Landsat / Copernicus.

Figure 8.1B River Amazon at Santarém, Brazil, where tidal range is about 20 cm. Photo is a low-season composite (probably December), and islands and sediment bars will be underwater by about March. The Amazon controls the level of the tributary Tapajós River, which forms a natural reservoir for about 100 km upstream from the confluence. The Santarém region hosted many indigenous settlements, and dark anthropogenic soils are common, generated by indigenous cultivation. Context for the area provided by Myrtle Shock. Maps data: Google, Maxar Technologies CNES/Airbus Landsat/Copernicus.

Much of South America is made up of ancient cratons that include the Guyana and Brazilian Highlands, but the bulk of the high Andes is young, dating back some 25 million years. As always, the rivers responded with panache to the changing landscape. The reversal of the Amazon from west to east and its breakthrough to the Atlantic is one of the great epics of Earth history. More than any other continent, South America's recent geological history is the story of one dominant river.

The river journey of Alexander von Humboldt

In 1800, Alexander von Humboldt and Aimé Bonpland, a good-humoured man whom Humboldt met in Paris carrying a satchel of plants, were overjoyed to paddle into the Casiquiare Canal (Fig. 8.2).[7] The 'canal', 300 km long, is a natural channel that connects the Río Negro, an Amazon tributary, with the Orinoco. European armchair scientists had declared such a connecting

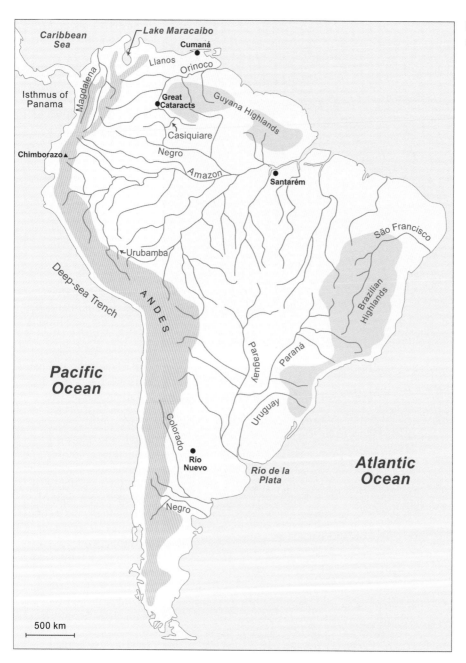

Figure 8.2 Modern rivers of South America.

waterway to be impossible, but the indigenous people and early European adventurers knew better. Humboldt fixed the canal's position carefully with his instruments before their dugout canoe nosed between walls of vegetation, carrying its menagerie of small jungle cats, monkeys, and a toucan.

The canal may be unique, a major river capture in progress as the Río Negro turns the upper Orinoco over to the Amazon. Although water flows between the rivers, the canal forms a barrier to the migration of freshwater organisms on account of the differing acidity and temperature of the black and white water.[8]

In the days that followed, Humboldt was less enthusiastic. The journey up the Orinoco had seemed straightforward in Europe, where he had written before embarking that he would explore the unity of nature. Now all nature seemed to have united against him. Unlike the black water and crystal sand of the Río Negro, the whitewater canal was plagued with torrential rain and a cloud of insects so unendurable that even Humboldt's meticulous diary trailed away. They were reduced to eating ants, and a jaguar ate their dog.

It had been an extraordinary journey, both scientifically and personally. As Chief Inspector in the Prussian Department of Mines, Humboldt had designed safety equipment and established a free school for miners, training them in geology, mining law, and river dynamics. However, his friendship with poet and scientist Johann von Goethe encouraged a holistic view of the Earth and reignited a childhood desire for travel. Impressed by Humboldt's mining reputation and the ability of a Prussian to speak Spanish, the King of Spain granted them permission to visit the Spanish-American colonies. Humboldt disembarked at Cumaná in Venezuela where, before landing, he conversed far into the night with an indigenous seafarer. He loathed the Cumaná slave market, but he showed local ladies their hair lice under the microscope.

South America was a sensory extravaganza for the eager travellers, but reaching the Casiquiaire was a tough challenge. They journeyed inland across the *Llanos*, a wide belt of small rivers and lakes where, in the wet season, the ranchers travelled the landscape by boat.[9] Humboldt and Bonpland, however, crossed the plains in the dry season when the temperature of the cracked soils soared to 50°C. Vampire bats hovered over their hammocks, and an alligator chased them from a waterhole. Hearing that they wished to study the potentially lethal electric eels, local people drove wild horses into a pond where they took intense shocks, exhausting the eels' electric charge. Even so, Humboldt experienced strong shocks as he experimented with the captured eels.

So wide was the Orinoco that the distant mountains seemed to spring from the water as the party canoed upriver to the Great Cataracts, the limit of scientific knowledge at that time. The travellers saw dolphins, piranhas, and stingrays. Gigantic anacondas swam alongside and herds of capybaras like huge guinea pigs paddled around the canoe. From the jungle emerged the jaguars, kept at bay through the night by campfires. Sloths hung from the branches, and the sand bars were laced with turtle eggs that were mined for oil by settlers and already growing scarce. The party slept on rocky islands among clouds of insects so dense that Humboldt was unable to see the sky through his astronomical instruments. He observed the making of curare poison and, accidentally spilling some on a sock, would have died horribly had he put the sock on a foot bleeding from insect bites.

Humboldt and Bonpland crossed the Andes, travelling by canoe up the Magdalena River. Ascending the volcano of Chimborazo, believed at the time to be the world's tallest mountain, they reached 19,286 feet (nearly 5900 m), the highest altitude yet attained. The climb made Humboldt famous and a bullfight was held in his honour. Mountains, he thought, were linked to violent volcanic upheavals, as Darwin later supposed. He documented an ocean current off Peru, later named for him, and visited Thomas Jefferson in Philadelphia, admiring the American political system and the President's pleasure in playing with his grandchildren but declaring his opposition to slavery. Returning to Europe with 60,000 plant specimens, Humboldt declared 'I was ever aware that one breath, from pole to pole, breathes one single life into rocks, plants and animals, and into the swelling breast of man'.[10] He settled in Paris and met a suspicious Napoleon who commented, 'So you're interested in botany? So is my wife', and later ordered the secret police to ransack his apartment. However, when Prussian troops occupied Paris in 1814, Humboldt intervened to spare the Natural History Museum.

Humboldt influenced generations of travellers. Charles Darwin attributed the direction of his life to Humboldt's writings, as did Alfred Russel Wallace, co-presenter of the theory of evolution with Darwin, who sailed up the Amazon in 1848.[11] Wallace reached his initial evolutionary views while living on the Río Negro, and he founded the science of biogeography by observing the distribution of the river fish (nearly half of all freshwater fish species on Earth live in tropical South America). Travelling up the river, Wallace passed the Casiquiaire Canal and exchanged regards with a jaguar that padded out of the jungle.

The Amazon changes its mind

More than 90 per cent of South America's modern river flow ends in the Atlantic and Caribbean, but when South America was part of Pangea, many of its rivers flowed west to the Pacific. Among the Pacific drainages was a hypothetical river called the *Sanozama*, or Amazonas reversed (Fig. 8.3).[12]

As South America and Africa drifted apart in the Cretaceous, the Paraná plume generated enormous

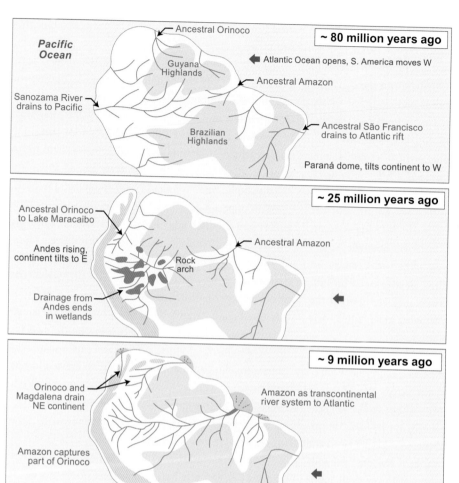

Figure 8.3 Stages in the evolution of the Amazon and other rivers in the northern part of South America. The 'Sanozama' river initially flowed west along the line of the modern Amazon, but drainage reversed to form the Amazon system as the Andes rose, generating a transcontinental river by nine million years ago. After Figueiredo et al., 2009.

volcanic eruptions and created rifts with uplifted flanks. Straddling the rifting continents was a dome from which rivers flowed into the interiors of Africa and of South America—the ancestral São Francisco, Paraná, and Uruguay rivers. With courses around cratons and long reaches parallel to the sea, the rivers were circuitous and inefficient. In contrast, a short blackwater river that would later become the downstream part of the Amazon followed a more direct eastward route along a rift.

Then, as South America drifted west over the Pacific, the Andes began to rise and the continent tilted to the east until, by the end of the Cretaceous, the rivers had reversed their flow into shallow inland seas or out to the Atlantic. By the end of the Oligocene 25 million years ago, sediment was pouring eastward from the Andes into wetlands that were closed off by a ridge of rock between the Guyana and Brazilian highlands.

Strata and fossils preserved from those times tell of a lazy landscape where dolphins and crocodiles inhabited the winding and switching rivers and anacondas, armadillos, and sloths frequented a maze of wetland creeks. A small ancestral Amazon east of the arch drained into the Atlantic.

As the Andes approached their present scale in the Miocene about 12 million years ago, the rivers overtopped the ridge and rift faults guided a connected Amazon to the Atlantic. By nine million years ago a river of transcontinental proportions was pouring its massive sediment load into the ocean.

Across a drainage divide to the south, the Paraguay crosses the continent from the Andes, picking up the Paraná and Uruguay in a course to the Atlantic as the Río de la Plata. The Paraguay features prominent meander belts with complex patterns of meander cutoffs (Fig. 8.4). Its tributaries include the Taquari

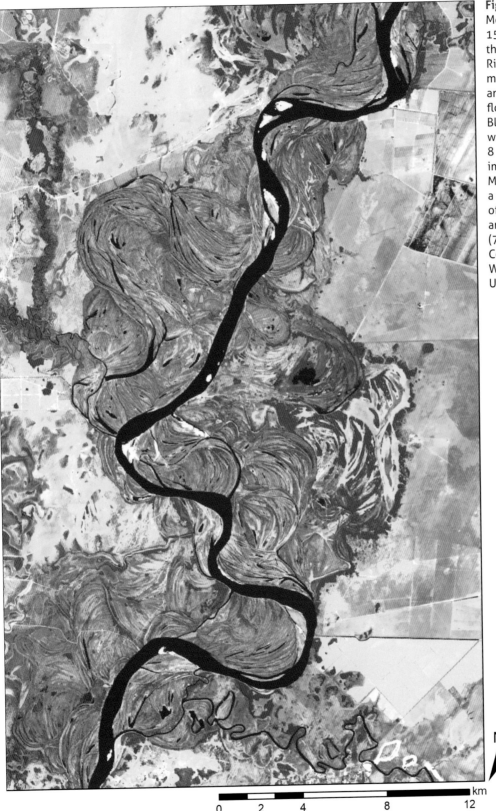

Figure 8.4
Meander-belt
15 km wide of
the Paraguay
River, with cutoff
meander lobes
and adjoining
floodplain.
Black areas are
water. Landsat
8 false-colour
image from 27
May 2020, using
a combination
of red, green
and blue bands
(7, 6 and 5).
Courtesy of Gary
Weissmann and
USGS.

N

km
0   2   4   8   12

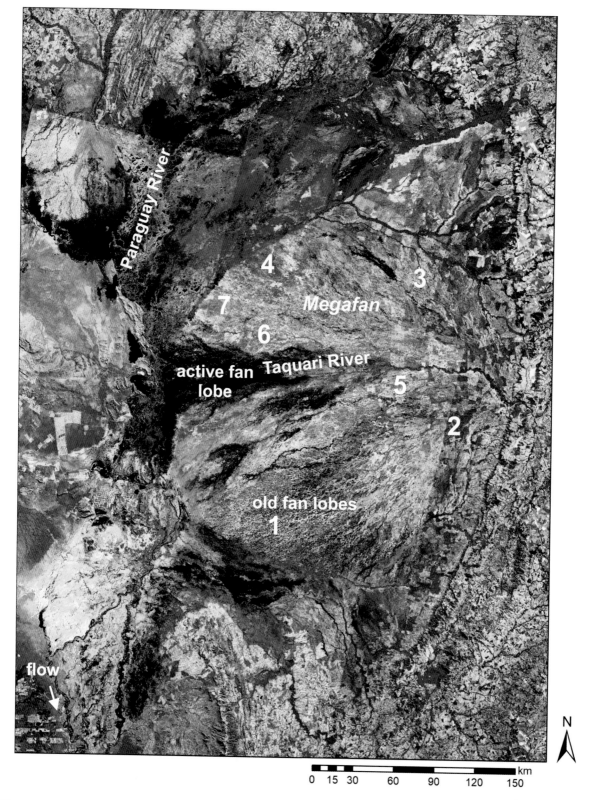

Figure 8.5 Taquari River and Megafan (250 km long) in the Pantanal wetlands, with inactive (numbered) and active fan lobes. The megafan drains westward to the Paraguay River. Black areas are water. Landsat 8 false-colour image from 2019, using a combination of red, green and blue bands (7, 6 and 5). Courtesy of Gary Weissmann and USGS.

River and megafan of the Pantanal wetlands, which has built westwards for 250 km as one of the largest megafans on Earth (Fig. 8.5). The megafan comprises the active modern lobe of the Taquari River and a series of older inactive lobes that reflect the repeated breakout of channels blocked with sediment and wood, cutting new courses through the sandy sediment.

Before the Amazon gathered strength, the Orinoco had dominated the continent's northern drainage, with an ancestral system in the Guyana Highlands that may date back to the Jurassic. The proto-Orinoco once flowed north through the area of Lake Maracaibo to the Caribbean, as indicated by freshwater vertebrate fossils that belong to the Orinoco system. Some 12 million years ago, rising mountain chains deflected the Orinoco eastward to the Atlantic and created a mountain valley for the Magdalena, both relatively young rivers in their modern courses. Thereafter, the Amazon captured much of the Orinoco drainage.

The electric fish of the Orinoco that shocked Humboldt resemble those of the Amazon headwaters but not those of the Amazon mainstem—a reminder of the days when modern Amazon tributaries were once part of the Orinoco. Incredibly, the world's largest river is less than ten million years old in its modern course.

Jack Gregory visits South America

In 1932 at the age of 68, Jack Gregory decided to visit Peru.[13] With his usual plethora of interests, he intended to determine the age of the Andes, collect lavas, examine the high deserts, view Inca ruins, and boat down the Amazon headwaters to Brazil. The party crossed the ranges on mules and, after being mistaken for bandits, embarked in canoes down the Urubamba. Landslides had narrowed the channel and intensified the flow. While traversing a precipitous gorge, Gregory's canoe turned over and he drowned in a whirlpool. As a friend noted, it was the kind of death that Gregory would have wanted.

CHAPTER 9

CANYONS AND CATARACTS IN NORTH AMERICA

Terrains of the imagination

In 1975, I rafted down the San Juan River, a tributary of the Colorado in Utah. Loaded with camping gear, the rubber rafts drifted in the turbulent eddies, past gnarled cottonwoods and flood layers of rippled sand. A hummingbird chattered in front of my face, retiring to a tiny woven nest on a branch. All around were red rock pinnacles like gigantic sculptures in an immense solitude, a brooding landscape of silent presences in the dawn mist.

I had recently arrived in North America, and the West and its rivers were an extraordinary experience. There was the Black Canyon of the Gunnison, so deep and narrow that, from the rim, the river far below lay obscured in gloom. At the Goosenecks, the San Juan had carved a deep, sinuous course through the rock, entrenching its original meanders (Fig. 9.1). Dry sandy creeks crossed the midwestern plains. And Native Americans had inscribed the desert varnish of the rock faces with antlered deer, humans on foot and horseback, and mysterious forms. Knowing the seasonal rivers and secluded springs, they had farmed the wild San Juan basin for thousands of years.

Back home as the Canadian winter descended, the river landscapes of the West continued to fill my mind. Rafting past cliffs of fluvial strata, I had seen the history of North America through the eyes of Pangea's long-extinct rivers. I had crossed modern rivers that arose with growing mountains, volcanoes, and rifts as the continent's drainage evolved (Fig. 9.2). I had followed the first inhabitants and the scientists who established river science. And, crossing the Mississippi, did I hear the whisper of steamboat pilot Mark Twain, speaking of the great river that he knew so well? In the words of San Juan river poet Ann Weiler Walka: 'In my room above a crowded, wintry street I have explored the mysterious border between a river's canyons and the terrain of my imagination.'[1]

Figure 9.1 The Goosenecks of Utah, where the San Juan River has incised meanders through Carboniferous strata. Volcanic neck in the distance.

Colorado River and the origins of river science

Lieutenant Joseph Christmas Ives stood on the lip of a mile-deep chasm (Fig. 9.3).[2] Below his feet a sheer wall of rock fell away to the tiny thread of the Colorado River far below. The canyon was so deep that the first Spanish adventurers in the 1540s believed that the river was only a few metres wide until, descending partway, they realized that each rock mass along the channel was taller than the cathedral of Seville. Beyond the chasm, a vast plateau of red and white rock stretched away as far as the eye could see.

Ives watched cloud shadows drift across the landscape, picking out the contours of the pinnacles. Along the cliff top the wind rustled the pinyon pines that clung to the ledges. Turkey vultures worked the air currents along the precipice, now below and now above him, and a raven sheltering among the pine branches stared up at him, its feathers rustling in the breeze.

Figure 9.2 Modern rivers of North America.

But Joseph Ives was not in a cheerful mood on that day in 1857. His curious lack of prescience in describing the Grand Canyon, now visited yearly by millions of tourists, seems astounding:

The region last explored is, of course, altogether valueless. It can be approached only from the south, and after entering it there is nothing to do but to leave. Ours has been the first, and will doubtless be the last, party of whites to visit this profitless locality. It seems intended by nature that the Colorado river, along the greater portion of its lonely and majestic way, shall be forever unvisited and undisturbed.[3]

But pity poor Joseph—it had been a trying time. To supply the United States Army in its skirmish with Mormon settlers, Ives had shipped sections of a paddlewheeler to the mouth of the Colorado River via an overland crossing of the Isthmus of Panama. The ship had been assembled in a pit and floated into the channel along a hand-dug canal. Sailing up the river, Ives had

Figure 9.3 The Grand Canyon in Arizona, where the Colorado River has cut through flat-lying Mesozoic and Paleozoic strata. The Great Unconformity between Cambrian strata and underlying dark Precambrian metamorphic rocks is visible low in the canyon. Much of the canyon was cut within the past six million years.

grappled with sand bars, rapids, and a broken rudder. Then, in the chasm of the Grand Canyon (now drowned behind the Hoover Dam) they ran onto the rocks. Ives was pitched into the bottom of the boat, the boiler was damaged, and the steam pipe was bent double.

With upstream navigation impossible, Ives proceeded with mules along a Mormon road towards Utah. He wrote of his experience:

> The extent and magnitude of the system of cañons in that direction is astounding. The plateau is cut into shreds by these gigantic chasms, and resembles a vast ruin. Belts of country miles in width have been swept away, leaving only isolated mountains standing in the gap. Fissures so profound that the eye cannot penetrate their depths are separated by walls whose thickness one can almost span, and slender spires that seem tottering upon their bases shoot up thousands of feet from the vaults below.[4]

At last the party stood on the rim of the Grand Canyon. But attempts to reach the Colorado River proved fruitless. The cliff trails were so sheer that the men had to crawl, overcome with vertigo, and the mules sidled along inches from a dizzying precipice. Ives could barely alight safely and he was fortunate to find an alcove where the mules could turn round.

John Strong Newberry, however, left Ives in no doubt that there lay before them 'the most splendid exposure of stratified rock that there is in the world'.[5] The geologist suggested that the canyon, accessible only to the 'winged bird', had been formed by the river because the strata matched on either side. No colossal rupture of the land had created a path for the flow. Such a notion, he declared, 'lacks a single requisite to acceptance, and that is *truth*'. The simple power of river water had formed the canyon, which had never brimmed with the raging waters of a universal flood.

But it was John Wesley Powell who put the Grand Canyon on the map.[6] Powell had worked on family farms in the Midwest, and then, in the words of biographer Donald Worster,

> Wes discovered rivers. They flowed through the landscape of his mind like songs of freedom and

escape. They sang of catfish, beaver, blue herons, grape vines festooning the trees, the smell of mud. He had known rivers since a boy … he began to turn back to rivers to find what no college offered, an education out of doors where on his own he could learn to read the book of nature.[7]

Powell rowed from St Anthony Falls at Minneapolis to New Orleans, the entire navigable length of the Mississippi. Enlisting in the Union Army, he became an artillery major during the Civil War. Raising his arm for a barrage at the Battle of Shiloh, Powell was hit in the wrist and the arm was later amputated below the elbow—an injury that troubled him all his life.

But the West was calling and in 1869 Powell's expedition embarked on a three-month journey down the Colorado River in small boats. The party ran rapids where the boats were swamped and the oars smashed, salvaging barometers and a whisky keg from a wrecked boat stranded in mid river. They crawled along ledges with ropes to lower the boats down waterfalls, and a flash flood chased Powell down a canyon. Climbing a precarious cliff, the one-armed veteran stuck below an overhang and was only saved when an expedition member used his own trousers as a makeshift rope to draw Powell up. He seemed able to manage without food or shelter as long as he could examine the rocks, the boatmen grumbled.[8]

The adventurers named Flaming Gorge with its brilliant red rocks, Canyon of Lodore after a turgid poem, and Whirlpool and Desolation canyons. They drifted through Glen Canyon with its tributary gorges and groves of ferns, where masonry houses and carved steps testified to former Native American occupation. Sounds reverberated sweetly in Music Temple. The narrow chasm of Marble Canyon below Vermilion Cliffs was desolate enough for any love-sick poet, an expedition member noted. At last they entered the Grand Canyon itself, likened by writer Wallace Stegner to a mountain range in a ditch.[9] The vicious rapids of Lava Falls, the jagged remnants of a lava dam, finally swept them through.

As Stegner notes, Powell's 1895 account made the most of the river drama, and it should be read as an adventure story.[10] But what an adventure story!

Powell knew the West like no other scientist of his day, and he dedicated the rest of his life to understand-

ing its river landscapes and the challenges of dryland settlement (see Chapter 20). He was fascinated by the indigenous peoples who had moved through the arid West more than 10,000 years earlier, founding large communities sustained by irrigated agriculture. As Powell knew well, they were the true explorers of the canyon country.

More than any other river, the Colorado has been the crucible of river science. Over the following decades, Powell and his associates G.K. (Grove Karl) Gilbert and Clarence Dutton revolutionized concepts of landscape. Gilbert had a mind for learning like a horse leech, his aunt said.[11] A rigorous scientist, he valued precise measurement (visiting the giant redwoods of California, he studied the geometric shape of a spider's web). In 1869 Newberry recommended him for a military expedition in the West, during which he observed the planet Venus from the gloom of a canyon floor. Gilbert and Powell shared an interest in cigars, cards, and billiards, the latter appealing to Gilbert's sense of geometry. Gilbert found Powell rich in ideas and free in communicating them, but Gilbert's assistant Bailey Willis thought that Powell gained at least as much in return. With his mechanistic mind, Gilbert revelled in the challenge laid down by the western rivers—the balance of forces to which they responded, the rates of landscape erosion, their adjustments over waterfalls. They operated in a dynamic equilibrium.

Clarence Dutton completed the triumvirate.[12] A polymath Civil War veteran with abilities in mathematics, literature, athletics, and oratory, Dutton became interested in geology while working at a military arsenal. Transferred to Texas late in his career, he contemplated writing a cowboy novel. In 1882, Dutton coined the term *isostasy* for vertical movements of the Earth's crust, recognizing that erosion leads to uplift of the land because gravity works to restore the balance. It was Dutton who penned some of the most imaginative early descriptions of the Colorado River:

At the foot of this palisade is a platform through which meanders the inner gorge, in whose dark and somber depths flows the river. Only in one place can the water surface be seen … Yet we know that it is a large river, a hundred and fifty yards wide, with a headlong torrent foaming and plunging over rocky rapids. … Not only are we deceived, but we are conscious that we are deceived, and yet we cannot conquer the deception … In all the vast space beneath and around us there is very little upon which the mind can linger restfully. It is completely filled with objects of gigantic size and amazing form, and as the mind wanders over them it is hopelessly bewildered and lost.[13]

Why and when had the river carved the Grand Canyon? The canyon plunged through the Kaibab limestone uplands: had the river existed before the uplands and maintained its course as the land rose—an *antecedent* river as Powell maintained (Fig. 6.6)? Or had the river drained a subdued landscape of younger strata, perhaps a dried-up lakebed, and later cut down into older strata with which it had no initial connection—a *superposed* river, as Dutton suggested? The river and the plateau: which was the chicken, and which was the egg? They defined *base level*, the surface of an ocean or lake that limits the river's ability to erode down, and the *graded river*, neatly balanced between the sediment available for transport and the ability of the river to transport it.

A fourth scientist waded into the debate—William Morris Davis, who toured the Grand Canyon with a pack train in 1900.[14] A towering figure in landform analysis, Davis came from tough Quaker abolitionist stock in Philadelphia: his grandmother Lucretia Mott had quietly awaited a pro-slavery mob intent on burning down their house (they eventually passed by). He knew Powell, Dutton and Gilbert personally and found their written accounts so vivid that visitors seemed already to know the region.

Davis puzzled over the unconformities in the canyon walls which, like James Hutton, he attributed to periods of uplift and river action. Why was the Kaibab limestone country riddled with sink holes and springs, a *karst landscape* that yielded evidence for a former underground drainage? And where in such an arid region had the water come from to cut a canyon that was 350 km long, 30 km wide, and two kilometres deep? It was too simple, he thought, to describe the river as antecedent or superposed. The Colorado might be a *subsequent* river that had followed later folds and faults. Musing on the extensive plateau, he added his own term *peneplain* for a landscape eroded almost flat by rivers.

And at what stage of such a complex history could we recognize the *Colorado River*? The Grand Canyon was a 'precocious young valley', a 'very respectable beginning, but nothing more.'[15] Davis noted that 'geography is only the geology of to-day; and we cannot properly appreciate to-day without some understanding of the many geological yesterdays.' Landform description, he suggested, should not be the domain only of geologists and poets.

Many aspects of the Colorado's history remain controversial.[16] The river has variously flowed east or west, depending on the ups and downs of the Cordillera and the plateau. In the words of folksinger Katie Lee, 'Pore Colly Raddy, yer just lookin' for a sea … wonder if yer mama ever told ya what sea it oughta be.'[17]

The Cordillera of North America began to rise in the Cretaceous after the opening Atlantic forced the continent westward over the Pacific crust.[18] Ancestral rivers that would have included an early Colorado River flowed east from the rising mountains to the shallow ocean of the Western Interior Seaway that periodically covered the Midwest, possibly at times draining to the Bell River of Canada and out to the Atlantic Ocean (see Chapter 10). Then, through the Oligocene and early Miocene, from about 30 to 16 million years ago, tectonic activity set the plateau rising and the rivers became ephemeral and confused, dying out in inland basins and windblown sand dunes. Some 16 million years ago in the Miocene, fault-bounded rifts of the Basin and Range extended and lowered the land to the west, and a more organized Colorado River followed a westward path within the plateau area, responding to the change in gradient. During that period, a western drainage probably breached the Kaibab uplands to connect with an eastern drainage. Like the Rhine's piracy of the Danube headwaters, the breach may have been aided by the underground limestone drainage that had puzzled Davis (Fig. 9.4), as caves and passages carried water under the divide until the overlying rock collapsed to form a gorge that the river occupied. As little as six million years ago in the latest Miocene, the river spilled westwards into a series of lake basins, and by four million years ago in the Pliocene, the river was connected to the Pacific. The lowering of the river's base-level to the ocean would have promoted rapid incision and canyon formation in its lower reaches, and the flow built a large braided-river plain and delta into the Gulf of California, aided

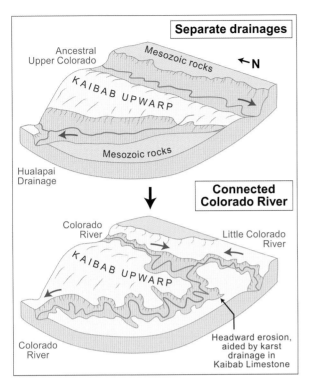

Figure 9.4 Stages in the evolution of the Colorado River. Development of a connected drainage system, sometime after 16 million years ago, was assisted by underground drainage through limestones of the Kaibab Uplift. After Ranney, 2012.

by abundant sediment from the recently integrated drainage system.

But had normal river flow over geological time really formed the Grand Canyon? The jagged lava flows that impeded Powell had once dammed a lake hundreds of metres deep and more than 100 km long within the canyon. When the dam collapsed, megafloods cut into the riverbed and swept boulders to levels far above the present channel. During the Ice Age, glacial meltwater swept the canyon. Even small lakes, dammed by rock falls in side valleys like those of Glen Canyon, would have released erosive floods.[19]

In 1882, Dutton wrote:

No doubt the question will often be asked, how long has been the time occupied in the excavation of the Grand Cañon? Unfortunately there is no mystery more inscrutable than the duration of geological time … In the Plateau Country Nature has, in some respects, been more communicative than in other regions, and has answered many questions far more fully and graciously. But here, as elsewhere, whenever we interrogate her about time other than relative, her lips are sternly closed, and her face becomes as the face of the Sphinx.[20]

With the advent of numerical dating, we now know that a connected Colorado River is remarkably young. Parts of the Grand Canyon system were formed tens of millions of years ago by unconnected river systems and during local drainage connection that was accompanied by base-level change. But the drainage integration that finally brought the Colorado River to the Pacific dates back only a few million years, promoting the incision of the western part of the Grand Canyon in record time.

The Mississippi wanderings of Mark Twain

Samuel Langhorne Clemens, better known as Mark Twain, grew up in the town of Hannibal on the Mississippi, the *Great River* in Ojibway.[21] In 1803–6, only three decades before Twain's birth, Meriwether Lewis and William Clark had documented many parts of the Mississippi system.[22] Lewis was a biologist and a writer with a fine prose style, a former aide to President Thomas Jefferson, whereas Clark was a riverman and cartographer who made the region's first accurate river maps and wrote cryptic notes. The expedition journeyed down the Ohio to the Mississippi, up the Missouri and, crossing the continental divide, followed the Snake and Columbia rivers to the Pacific. Often drawing on information from Native Americans, they measured channel widths and meander dimensions, and they documented river terraces high above the modern channels. They wrote about ice jams and beaver dams. The abundance of dry channels was a surprise to the two men, who were used to the perennially flowing rivers of Appalachia.

Twain's parents moved west with the flood of settlement that followed the expedition, but they proved exceptionally gifted at exploring commercial backwaters. As Twain put it, 'It is good to begin life poor; it is good to begin life rich—these are wholesome; but to begin it poor and *prospectively* rich! The man who has not experienced it cannot imagine the curse of it'.[23] A sickly tearaway, Twain once narrowly escaped drowning by jumping across ice floes when the frozen Mississippi broke up. His mother confessed that she was constantly uneasy about him. Afraid I wouldn't live? asked Twain. 'No—afraid you would.'[24]

Hannibal was a river port, alive with lumber rafts and shipping, and every child's admiration was fixed on the romantic steamboats with their paddle wheels and famous pilots. But Twain was also acquainted with the dark side of river life—death by drowning, cholera, murder, and the slave trade. At a young age he could remember everything, whether it had happened or not. He planned an expedition to the Amazon, and when a $50 bill blew his way on a windy street, he took the money out of danger by purchasing a steamboat ticket south. Finding that no ships were departing from New Orleans for the Amazon anytime in the century, he apprenticed with steamboat pilot Horace Bixby.

The years on the Mississippi were the making of Twain. Bixby taught him the river from St Louis to New Orleans (Fig. 9.5), 2000 km of meanders and sand bars. At times the river was black with dead logs and fallen trees, and he spent hours sounding for the channel in drifting ice. In his words,

> piloting becomes another matter when you apply it to vast streams like the Mississippi and Missouri, whose alluvial banks cave and change constantly, whose snags are always hunting up new quarters, whose sand bars are never at rest, whose channels are forever dodging and shirking, and whose obstructions must be confronted in all nights and all weathers ... The face of the water, in time, became a wonderful book—a book that was a dead language to the uneducated passenger, but which told its mind to me without reserve, delivering its most cherished secrets as clearly as if it uttered them with a voice.[25]

Twain's river schooling brought him into contact with pimps, prostitutes, gamblers, and a pilot who quoted Shakespeare between orders to the crew. And there was the mate who mixed profanity with geological terms from Charles Lyell's textbook, violently castigating the deckhands as 'Old Silurian Invertebrates out of the Incandescent Anisodactylous Post-Pliocene Period'.[26] Steaming upriver, Twain may have passed the teenage John Wesley Powell rowing downstream.

At the onset of the Civil War, Twain sailed upriver and, as he expressed it, was shot through the chimneys. He joined a Confederate militia but resigned after learning more about retreating than the man who invented retreating.[27] He worked as a silver miner in Nevada and, moving to San Francisco, became a 'scribbler of books'. Later journeys along the Mississippi brought back youthful memories and contributed to *Life on the Mississippi* and *Adventures of Huckleberry Finn*. He adopted the pseudonym *Mark Twain* from the calls of the leadsmen sounding ahead of the big steamboats—twelve feet of water at the two-fathom mark, a suggestion of danger. The river gave him a name.

What did the Mississippi mean to Twain? Commercial mastery of the river took away grace, beauty, and poetry: a log floating in the sunset meant that the river was rising, a slanting mark on the water denoted a sand bar that could destroy a steamboat. But powerfully and authentically through *Adventures of Huckleberry Finn* rolls the Great River, a refuge and a friend, bearing Huck and Jim on their raft away from the violence and self-interest that Twain came to abhor. Perhaps it was only through the eyes of young Huck Finn that he could describe the river as he really knew it:

> It was a monstrous big river down there—sometimes a mile and a half wide; we run nights, and laid up and hid day-times; as soon as night was most gone, we stopped navigating and tied up ... Not a sound, anywhere—perfectly still—just like the whole world was asleep, only sometimes the bull-frogs a-cluttering, maybe. The first thing to see, looking away over the water, was a kind of dull line—that was the woods on t'other side—you couldn't make nothing else out ... then the river

Figure 9.5 The Mississippi River and Delta south of New Orleans, with prominent natural levees and salt marshes with distributary channels.

softened up, away off, and warn't black any more, but gray; you could see little dark spots drifting along, ever so far away—trading scows, and such things; and long black streaks—rafts; sometimes you could hear a sweep screaking; or jumbled up voices, it was so still, and sounds come so far; and by-and by you could see a streak on the water which you know by the look of the streak that there's a snag there in a swift current which breaks on it and makes that streak look that way; and you see the mist curl up off of the water, and the east reddens up, and the river, and you make out a log cabin in the edge of the woods, away on the bank on t'other side of the river … and next you've got the full day, and everything smiling in the sun, and the song-birds just going it![28]

Later in life, nightmares had Twain guiding a steamboat into a shadow that might be a bluff, an island, or the black wall of night.[29] Nevertheless, he loved to remember the splendid river shimmering in the light. He died in 1910, his birth and death coinciding with Halley's Comet, both 'unaccountable freaks' who had come in together and must go out together.[30]

It is possible that no living river on Earth follows a course as ancient as that of the Mississippi.[31] A river has flowed south along its present course with only occasional interruptions for more than 300 million years, when a Carboniferous river as large as the Amazon ran west of the Appalachian Mountains, at that time of Andean scale although substantially cut down by erosion through the Permian and Triassic. Fault activity during the Triassic may have interrupted coherent drainage for a while. But for more than 150 million years since a short-lived marine flooding in the Jurassic, the Great River has run to the Gulf of Mexico along a rift zone that developed above a failed Proterozoic rift at depth. The faults are still active, and earthquakes in the Mississippi Valley near New Madrid in the early 1800s liquefied the river sand, which erupted up through the surface clay to build hundreds of tiny sand volcanoes.[32]

But the Mississippi had competition for access to the Gulf of Mexico, which formed as North America began to drift west in the Jurassic.[33] Cretaceous rivers flowed northwest from the Appalachian Mountains to the Boreal Sea through a catchment that included much of the continent, but the continent's drainage was dramatically reorganized as the Cordillera began to rise

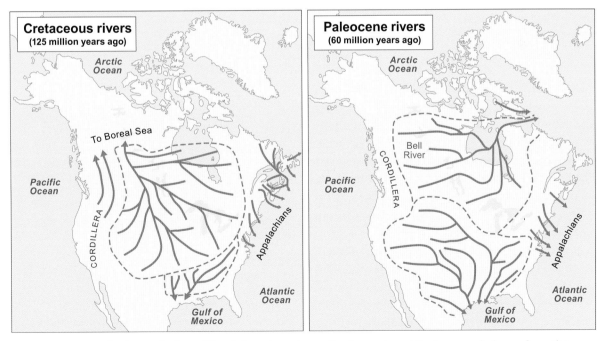

Figure 9.6 Stages in the evolution of North American rivers. North-westward Cretaceous drainage from the Appalachian Mountains was progressively replaced by eastward drainage from the rising Cordillera from the Paleocene onwards, with an increasing influx to the Gulf of Mexico. After Blum and Pecha, 2014.

and the Western Interior Seaway developed (Fig. 9.6), bringing to the Gulf a suite of rivers that included the Rio Bravo, Rio Grande, Guadalupe, another Colorado, Red, and Missouri-Mississippi. Their courses and discharge were strongly affected by tectonic events, volcanic activity, climate change, and river capture. By the Paleocene, a flood of Cordilleran sediment was pouring into the Gulf, where the Colorado and Mississippi built large deltas. Further north, the Bell River flowed to the Atlantic from the rising Canadian Cordillera. The Mississippi dwindled through the Eocene and may have died out completely for a while as western volcanoes erupted and the ancestral Platte and Arkansas carved courses through hundreds of metres of ash, at times outcompeting the Mississippi. As the land rose and tilted eastward, enormous fans of river gravel and windblown sand built the Miocene Ogallala strata of the High Plains, strengthening the axial Mississippi system (Fig. 9.7). An ancestral Rio Grande flowed along a Miocene rift but failed at first to reach the Gulf. Eventually, renewed uplift in the Appalachians revitalized the Mississippi's eastern headwaters, and the river extended its catchment northwards, capturing the Red and Tennessee rivers. Pleistocene ice sheets routed meltwater floods down the Mississippi to the Gulf of Mexico and, entrenched by meltwater incision, the Mississippi consolidated its domination of the Gulf drainage.

The misfortunes of the Columbia River

Never was a river so buffeted by ill fortune as the Columbia. Rising in Canada, the Columbia flows north as a magnificent anastomosing stream that winds and divides through wetlands with beaver dams (Fig. 9.8). The river then swings south to the Pacific through the lavas of the Columbia Basalt Plateau and the Cascade Range, where Lewis and Clark mapped dramatic waterfalls and rock-cut channels.

Little is known of the Columbia before the Miocene 17 million years ago.[34] Then the westward-drifting continent overran a mantle plume at Yellowstone and explosive eruptions broke out. The Columbia Basalt Plateau rapidly built up nearly two kilometres of lava, which swept the river valley with incandescent torrents. With its channels obliterated, the Columbia cut canyons through the basalt, only to see them filled by more lava.

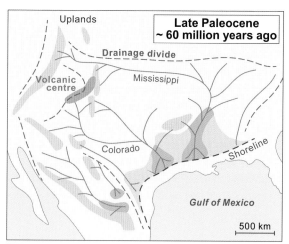

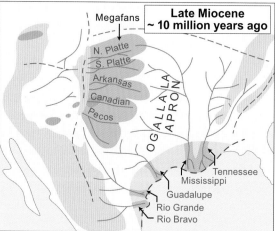

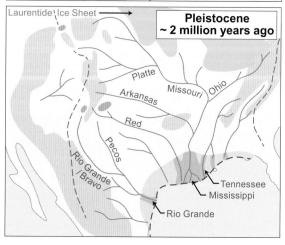

Figure 9.7 Stages in the evolution of rivers draining into the Gulf of Mexico from the Paleocene onwards. Tectonic and volcanic activity in the Cordillera, rejuvenation of the Appalachians, and the southward advance of the Laurentide Ice Sheet caused individual rivers to vary in importance. After Galloway et al., 2011.

Figure 9.8 Columbia River at Golden, British Columbia, Canada. In this reach, the river flows north with an anastomosing pattern through wetlands, with channel avulsion frequently caused by blockage due to beaver dams.

Figure 9.9 Great Falls on the Potomac River near Washington, D.C., where the river emerges from the Appalachian uplands. The latest phase of incision began about 37,000 years ago.

When the Cascade Range volcanoes erupted to the west above an east-dipping subduction zone, lava and volcanic debris flows pinned the struggling river against the basalt plateau to the east. The debris flows brought down gingko trees, entombed in lava and exquisitely petrified in silica at Gingko Petrified Forest State Park.

To the south the Snake River was also buffeted by the Yellowstone plume. At first the drainage radiated from the rising dome and the river flowed south. By the Pliocene four million years ago, the Snake had reversed its flow north and was captured by the Columbia, connecting the fish faunas.

The last straw for the Columbia came after the Ice Age. Gigantic meltwater floods swept the river as recently as 13,000 years ago, among the largest outburst floods ever on Earth (see Chapter 12). And near Mount St Helens the Columbia picks up a tributary that, during the 1981 eruption, discharged for a while more than 250,000 cubic metres a second of water, volcanic debris, and logs, approaching the volume of the Amazon in flood.[35]

Rivers of the Atlantic coast

The little Sackville River of eastern Canada, one of many rivers that drain the Appalachians, is in flood as the snowmelt releases the winter planet from its icy prison. Whirlpools surround sunken rocks, and the soft deep snow is fading from the forest, exposing yellow grass and rodent burrows. A faint young sun breaks through the mist. The little urban river, only 40 km long, seems surprisingly alive in the quiet backwater of winter as dog walkers brave the icy path and gulls drift over from the landfill. In summer, salmon still run up the Sackville River as they did in huge numbers before European settlement, and in 1936, baseball hero Babe Ruth took a fishing trip to a nearby stream and caught nine salmon before falling in.

Since the Jurassic, Canadian rivers like the Sackville have flowed to the Atlantic here.[36] As the continents drifted apart, rivers flowed south from Labrador and the Canadian Shield into a widening rift. Here and there the river sediments collapsed into gypsum sink holes and caves in a wide peneplain on the rift margin, carrying with them the first pine trees, turned to charcoal in Cretaceous forest fires.

More famous rivers rise in the Appalachian Mountains to the south—the Hudson flowing past New York, the Susquehanna as the longest river on the east coast, and the Potomac, which roars over Great Falls (Fig. 9.9) before flowing sedately past the White House in Washington DC. These are also Jurassic rivers that have broadly followed their ancestral courses to the opening Atlantic for the past 170 million years, building large Cretaceous deltas that match the present river positions.[37] Over the past 30 million years since the Oligocene and Miocene, these and other rivers have extended their catchments into softer rocks in the western Appalachians, carrying a flood of sediment offshore. I have travelled the world in search of the oldest rivers, but some of the oldest were in my backyard.

A historic river notebook

Forty years after rafting down the San Juan River, I join a field excursion of a different type to the historical archives of Philadelphia. Laid out by the curators for the assembled antiquarians—enthusiasts by any standard—is an extraordinary array of items. I gaze in amazement at Franklin's equipment for studying static electricity and peer at fossil bones collected for Thomas Jefferson from sites along the Ohio River. There are first editions of Galileo's *Dialogo* from 1632, Newton's *Principia* from 1687, and a flyleaf for *On the Origin of Species* in the author's handwriting.

On one of the cabinets lies a surveyor's notebook bound in red leather. I gently turn pages filled with a cursive script, the ink blotted here and there (Fig. 9.10). For 29th June Friday, I can just make out:

> the waters of the Kansas is verry disigreeably tasted to me … passed a large Island on the S. Side, opposit a large Sand bar, the Boat turned and was within Six Inches of Strikeing … the rapidity with which the Boat turned was so great that if her bow had Struck the Snag, She must have either turned over or the bow nocked off …[38]

I am holding one of the earliest accounts of river science, an 1804 notebook from the Lewis and Clark expedition. Replete with curious spellings and eccentric capitalization is the distinctive hand of riverman William Clark.

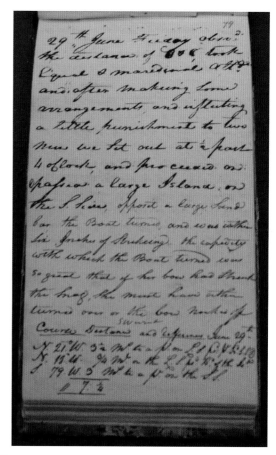

Figure 9.10 Page of a notebook from the Lewis and Clark expedition of 1803–6, describing the Kansas River, a tributary of the Missouri between the Platte and Arkansas rivers. Courtesy of the American Philosophical Society, Philadelphia, PA.

CHAPTER 10

A CANADIAN AMAZON

A plateau on the prairies

The Cypress Hills of southern Canada form a plateau between the Rocky Mountains and the Atlantic. Only a few hundred metres above the prairies, the hills divide the continent's rivers, sending them north to Hudson Bay or south to the Gulf of Mexico. A remnant of the original prairie grasslands, the plateau is carpeted with waist-high fescue grass and flowers. Dragonflies hover overhead and birds are everywhere—the grassland meadowlarks and bobolinks, mountain bluebirds, and the drifting menace of a harrier. The hillslopes are covered with lodgepole pines, and the First Nations hunters who followed the migrating buffalo herds constructed lodges and tipis from the pine trunks, sheltering in the forests through the bitter winters.[1] An occasional buffalo strayed into the woods, stragglers from the enormous herds that once numbered 60 million animals but, by the 1870s, were almost gone.

Capping the plateau is a layer of conglomerate, exposed in cliffs and drainages (Fig. 10.1A).[2] Smoothed as they tumbled along an ancient braided channel, the pebbles and boulders of quartzite rock were transported far to the east from the Rocky Mountains. Each pebble has a life story that is hundreds of millions of years old—the ripple marks of an ancestral ocean, and circular percussion marks where the torrent smashed the pebbles together (Fig. 10.1B). With the pebbles are the bones and teeth of crocodiles, turtles, and fish that lived in the river, along with the relics of river visitors—giant pigs, moles, snakes, and the crumbling yellowed bones of gigantic brontotheres, extinct rhino-like mammals. The youngest fossils date to the Miocene, some ten million years ago.

Geologists discovered the conglomerate in the late 1800s as they clambered up the Cypress Hills, hoping to find coal for the newly constructed Canadian Pacific Railway.[3] Instead they discovered relics of the Bell River, a long-forgotten river of continental scale.

Figure 10.1A Conglomerates of Oligocene age on the escarpment of the Cypress Hills, central Canada, in a 5m cliff. The gravel particles were transported hundreds of kilometres westwards from the Cordillera by the Bell River system that drained to the Atlantic Ocean.

Figure 10.1B Quartzite pebble with percussion marks from impacts during transport from the Cordillera. Coin is 2.4 cm in diameter.

The Bell River

The first account of the extinct river was published by Robert Bell in the Ottawa Evening Journal on May 17th of 1895.[4] Costing two cents and in its tenth year of production, the newspaper upheld the prestige of a capital city that until recently had been the wild logging settlement of Bytown. Bicycles were all the rage, and a bicycle club had fielded its second outing ('enterprising and pushing sort of fellows'). But Ottawa was still at heart a provincial town where a man could be fined for allowing his cow to wander the streets.

Of interest to the journal's readership that May was the London trial and conviction of Oscar Wilde. Many readers would have remembered his 1882 lecture on *The Decorative Arts*, part of a North American lecture tour in which he encountered floods, heat, and universal abuse.[5] He dined with Prime Minister Sir John A. Macdonald, whom he described (sighing) as a man of the world. Resplendent in a black velvet costume, he received more attention from parliamentary spectators with opera glasses than did the politicians.

In 1895, Robert Bell, for whom the extinct river was later named, was at the height of his powers.[6] His father was a friend of Geological Survey director William Logan, and the young Bell had worked as a geological assistant. During the next half century, Bell examined more of Canada by ship, canoe, and wagon than any other person of his generation. A strongly featured man with piercing eyes and a ferocious beard and whiskers, he is known for his studies of the tarsands of western Canada.

Bell knew the southern prairies in the days before they became the breadbasket of Canada. As his assistant wrote:

> We had two buckboards and two horses. Bell drove one and I the other … A vast solitude possessed the landscape, which, though ever shifting, was ever the same—an endless rolling surface of waving grass and myriads of flowers beneath a bright blue sky, with drifting snow-white clouds. Here and there the poplar bluffs and the gleam of saline waters in small lakes accentuated the glorious monotony of the prairie.[7]

The Ottawa Evening Journal reported his discovery under the heading of 'A Pre-glacial River—Dr. Bell of the Survey Tells the Royal Society of One—IT WAS IN NORTHERN CANADA—A LITTLE TIME BACK'. An enormous river (Bell claimed) had once drained the Canadian north when the land stood higher than at present before the glaciers eroded it down. He envisioned a vast drainage network from the Rocky Mountains to Labrador, reaching the Atlantic through the deep former valley of Hudson Strait (Fig. 10.2). The river would have been larger than the Amazon,

> of gigantic proportions compared with any river of the present world … The former existence of this

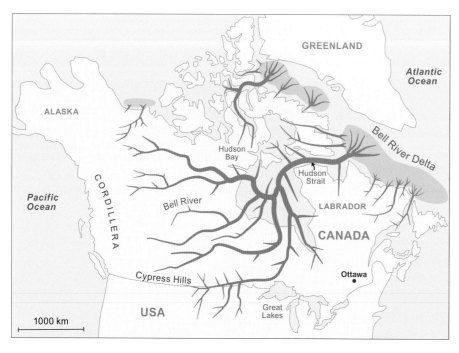

Figure 10.2 Map to show the drainage basin of the former Bell River, which was eliminated by the late Cenozoic ice sheets. After McMillan and Duk-Rodkin, 1995.

great river was not a mere speculation as to what might have been, but a necessary consequence of the elevation and change in the slope of the land, and it was proved in detail by a multitude of concordant facts all over the territory involved.[8]

We can imagine Robert Bell sitting by his campfire on a northern river, puzzling over the curious topography of the 'Canadian Oldland'. There was the Cypress Hills with its elevated gravel, surely laid down by a huge river. Ice sheets had once covered and eroded the continent, as suggested by Louis Agassiz half a century earlier. And there was Hudson Bay, now dividing the drainage but so shallow that even a slight rise in land level would expose the floor (Bell had accompanied naval expeditions to the bay in the 1880s). What was the Canadian landscape like before the Ice Age?

Some evening at sunset as he swatted away mosquitoes and puffed on his pipe (did he smoke a pipe?) it must all have come clear to him. The sun dips below the rim of the Canadian Shield and his understanding is paradoxically illuminated. Bell exclaims in wonder and accidentally snaps his pipe, sending embers cascading over his fingers and startling the ravens. Cursing as he brushes the glowing coals away, he inwardly exults at this eureka moment.

New evidence of a trans-continental river from Labrador

For the next 80 years, Bell's river was largely forgotten. Then startling new information came from oil wells off the Labrador coast.[9] Below the ocean lay eight kilometres of sandstone and mudstone, far too much to have come from the little rivers of the nearby coastal mountains. Much of the mud was uniform in composition, as ex-

pected for mud transported by a continent-wide river, and the clay yielded fossilized spores and pollen that could be traced to the rocks of western Canada. Just as the Amazon carries sediment from the Andes to the Atlantic, so the Bell River once brought Cordilleran sediment to the eastern ocean. Map reconstructions suggest a river basin nearly seven million square kilometres in area—larger than the Amazon, as Bell had suggested.[10]

A huge Cretaceous river system with its headwaters in the Appalachians had transported water and sediment northwestward (Fig. 9.6).[11] But this formidable river was rerouted as the Cordillera rose, and in its place the Bell River transported gravel, sand, and mud down a continent-wide slope to the Atlantic Ocean, reaching the sea through a breach in the coastal uplands. Then tectonic activity waned, and the Cordillera began to erode down. As the land rose through isostasy, the Bell River cut into its own sediment and then into the tough bedrock of the Canadian Shield below, leaving a remnant in the Cypress Hills above a peneplain cut by lateral river planation. No humans saw this amazing river. Only the brontothere herds knew it, coming down to the riverbank at dusk to drink.

And then ice sheets covered the northern landscapes. A river that moved untold volumes of water to the sea for some 50 million years passed out of existence almost overnight. As the most-recent ice sheets melted and sea level rose, shorter unconnected channels ended at Hudson Bay—the Saskatchewan, Nelson, Churchill, Peace, and Athabasca, all studied by Bell. No longer forced down by the weight of ice, the land around Hudson Bay is rising in places at more than a metre a century.[12] Thousands of years from now, the Bell River may once again flow to the Atlantic over the land that Robert Bell traversed by canoe and wagon. It is not clear who will be there to see it.

PART 3: HOW THE ICE AGE CHANGED RIVERS

CHAPTER 11

FROZEN OUT: NORTHERN RIVERS SCULPTED BY ICE

Gold on the Klondike River

In the summer of 1896, a group of gold prospectors wandered up the Klondike River, a tributary of the Yukon.[1] They included Keish, Shaaw Tláa, and Kaá Goox, members of the Carcross-Tagish First Nation, and a relative, George Washington Carmack, whose father had joined the 1849 California gold rush. George lived off the land and had little interest in gold. His cabin contained an organ and issues of *Scientific American*, and he wrote poetry in his spare time.

The group panned for gold as Carmack intoned 'To be or not to be'. It was to be. There in the pan was a yellow strand, four dollars of gold when a ten-cent pan was profitable. Trapped in bedrock crevices nearby, gold nuggets glittered in the sun. They performed a wild dance and Carmack hacked a blaze on a spruce tree, staking a claim to 500 feet of creek and floodplain.

The Klondike tributaries held some of the richest gold occurrences ever found, yielding more than 300 tons of the precious metal. As news of the discovery spread, men rushed to the Klondike (Fig. 11.1), and the tent encampment of Dawson City sprang up at the confluence of the Yukon and Klondike rivers (Fig. 11.2). Every claim on Eldorado Creek was worth half a million dollars or more in 1890s currency. Charley Anderson, known as 'the Lucky Swede', bought an untested claim during a drinking binge and, failing to regain his money, set out gloomily for a stretch of creek that netted him a million dollars. Thomas Lippy cashed in a claim high in the creeks for a lower claim because he needed timber for a cabin. His new claim yielded a million and a half dollars, his former claim yielded little.

The largest nuggets, some weighing nearly four kilograms, lay trapped on the bedrock under metres of sediment. To reach them, the miners thawed shafts through the frozen ground, hauling the muck to the surface to be washed during the spring melt. When his

Figure 11.1 The Klondike Gold Rush, from an 1898 engraving. Shutterstock from Everett Collection.

Figure 11.2 Confluence of the Klondike River (back left) and the Yukon River (foreground) at Dawson City, out of view to the left. The White Channel Gravel runs along parts of the hillside above the Klondike River. Shutterstock by Josef Hanus.

shaft reached the rock, Louis Rhodes lowered a candle and saw the nuggets winking up at him. People pulled nuggets from the frozen mounds when they needed small change.

In the following spring as the ice broke up on the Yukon River, sternwheelers with reinforced decks brought the first gold to Seattle and San Francisco. Down the gangplanks strode bearded miners bowed under the weight of gold dust and nuggets carried in caribou hides, blankets, suitcases, and jam jars. Seattle went wild. So many people rushed to the diggings that the city could barely function, and cyclists from Boston and New York pedalled furiously for the Klondike.

But the would-be miners had reckoned without the short Arctic summer and the long winter. Some got through by climbing ice cliffs on the Chilkoot Pass and traversing the White Pass with its nightmare litter of dead horses and abandoned equipment. Many foundered in the northern rivers. In the spring of 1898, 7000 newly constructed boats braved the Yukon rapids and whirlpools to reach Dawson City. An enterprising trader brought a crate of live chickens, selling the first egg for five dollars, and a boy made a fortune with newspapers. Lonely miners bought kittens for an ounce of gold, and a drunken donkey called Gerry the Bum cadged drinks in the bars.

For a while the city settled into a currency based on gold. Bartenders panned their fingernails at night and waitresses transferred gold dust on wet hands to secret pockets. The dust filtered between the floorboards, to be gathered when the buildings were demolished. And then within three years, it was all over. Gold was discovered at Nome downstream on the Yukon, and Dawson City emptied out.

Where had the gold come from? Although the nearby bedrock contains some gold-bearing veins, most of the veins were eroded and the gold was transported by the rivers.[2] Enterprising miners who knew their geology thawed down the terraces high above Bonanza Creek. In the White Channel Gravel, laid down on an erosional surface cut into the bedrock (a *strath terrace*), they found abundant nuggets which, over time, rolled down to the creeks below.

For tens of millions of years, the ancestral Yukon and its tributary the Klondike flowed south to the Pacific (Fig. 11.3).[3] Across a drainage divide to the north, other rivers flowed to the Bering Strait and the Arctic Ocean, and to the east the Bell River began its transcontinental journey to the Atlantic. When the first ice sheet advanced across the northern Cordillera in the latest Pliocene about 2.6 million years ago, ice blocked the Yukon River and

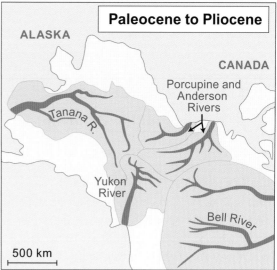
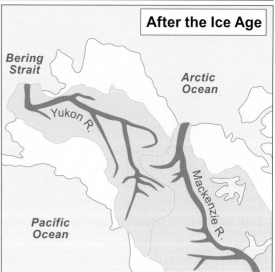

Figure 11.3 Changes in drainage associated with the Ice Age. The Yukon River initially flowed to the Pacific but later reversed its drainage to the Bering Strait. The Mackenzie River follows a meltwater valley along the mountain front to the Arctic Ocean, cutting through the former Bell River drainage, and may be as little as 12,000 years old. After Duk-Rodkin et al., 2001.

turned its flow north. Rising only 50 km from the Pacific, the river now runs for more than 3000 km to the Bering Strait, and the White Channel Gravel of the Klondike terraces was deposited after the flow reversed. Beyond the mountains gold largely drops out of the Yukon, which carries gravel to the coast like a giant conveyor belt, neither accumulating sediment nor cutting down. For several million years the Yukon has been G.K. Gilbert's rare equilibrium species—a *graded river*.

The Bell River was dismembered when, late in the last ice-sheet advance, a tongue of ice blocked the river's eastward flow and meltwater cut a deep valley northward along the mountain front. The valley is now occupied by the Mackenzie River, which has no ancestor along its present course and may be as little as 12,000 years old. Within the mountains, the Fraser River flowed north until meltwater erosion reversed its flow about 750,000 years ago, cutting the kilometre-deep Fraser Canyon en route to Vancouver, where it builds the largest delta on the west coast.

In 2016, the Yukon returned some of its flow to the Pacific. Meltwater from a retreating glacier cut an ice-walled canyon that, in a moment of geological time, diverted the Slims River, a Yukon tributary, to the south. Visitors found only an abandoned floodplain swept by dust storms.[4]

Louis Agassiz discovers the Ice Age

In 1837, the young Swiss geologist Louis Agassiz, whose name is indelibly linked to the discovery of the Ice Age, was slated to present his research on fossil fish at a conference in Switzerland.[5] Instead, Agassiz discussed boulders covered with scratches or *striations* in the nearby Jura Mountains, asserting that the boulders came from the Alps and that the striations had formed as ice ground the boulders against each other. Erratic boulders on landscapes were generally attributed to Noah's flood or, as Charles Lyell had suggested, to debris-covered icebergs. But Agassiz inferred that ice had once extended from the North Pole to the Mediterranean. There had been a geological epoch of ice.

There was pandemonium. Arguments among the incredulous scientists became so violent that other presentations were cancelled. And an ensuing field trip did little to convince the enraged scientists, some of whom muttered and glared as Agassiz walked on ahead.

Agassiz had reached his revolutionary understanding after touring mountain glaciers with an old friend, Jean de Charpentier. Agassiz gradually accepted the view of de Charpentier and earlier thinkers that the glaciers had once been more extensive. But not for the first time, villagers were ahead of the scientists.[6] De Charpentier had met a woodcutter who stated without hesitation that the Jura erratic boulders came from a distant glacier: the rocks were not from the local area

and floods could not have carried them to such high ground. De Charpentier gave him money to drink to the ancient glacier.

Agassiz realized that evidence of former ice extended across Europe, far beyond the range of Alpine glaciers. But there were no known analogues for such widespread ice, for the Greenland and Antarctic ice sheets were not recognized until later in the century. Eminent scientists were skeptical about an Ice Age, including eccentric professor William Buckland, who had visited the Alps with Agassiz but attributed the erratics to Noah's Flood. But a turning point came in 1840 when Agassiz, Buckland, and Murchison travelled across Scotland, the latter occasionally preceded by a highlander playing the bagpipes.[7] They found ice-worn rocks and boulders everywhere without glaciers to explain them, more evidence of what Agassiz called 'God's great plough'.[8] Murchison was not convinced. 'Agassiz gave us a great field-day on the Glaciers,' he wrote, 'and I think we shall end in having a compromise between himself and us of the floating icebergs!' But Buckland became a passionate advocate of ice sheets and, as Murchison noted: 'If you have not been frost-bitten by Buckland, you have at all events had plenty of friction, scratching, and polishing'.[9]

In 1846 Agassiz sailed to North America. Docking at Halifax in eastern Canada, he rushed from the ship and found familiar striations on the nearby rock outcrops. The Ice Age had not been confined to Europe.[10]

Ice and rivers

If huge volumes of ice had covered the land, the water must have come from the ocean. Writing in the *American Journal of Science* in 1842, Scottish journalist Charles Maclaren described the striated rock surfaces that he had observed with Agassiz, and then launched into a question that Agassiz had not considered.

> If we suppose the region from the 35th parallel to the north pole to be invested with a coat of ice thick enough to reach the summits of Jura … it is evident that the abstraction of such a quantity of water from the ocean would materially affect its depth … We find that the abstraction of the water necessary to form the said coat of ice would depress the ocean about 800 feet. Admitting

further, that one-eighth of the fluid yet remains locked up in the existing polar ices, it follows that the dissolution of the portion which has disappeared would raise the ocean nearly 700 feet. The only very uncertain element here is the depth of the ice; but even if this should be reduced one-half, we would still have an agent capable of producing a change of 350 feet on the level of the sea.[11]

Maclaren's back-of-the-envelope calculation was astute. Recent estimates put sea-level fall during the recent ice advance at about 120 m.[12]

Another Scotsman made a remarkable contribution to the question of ice ages.[13] James Croll had worked as a millwright and life-insurance salesman, but in 1857 his wife fell ill and Croll took her to Glasgow, where he found work as a museum janitor. The wages were low but, gratifyingly, he had access to the library. Fascinated by the Ice Age, Croll calculated how much energy the Earth received during its eccentric orbit around the Sun, demonstrating that the amount varied systematically and was sufficient to plunge the Earth periodically into an Ice Age.

Serbian scientist Milutin Milanković took Croll's work forward after celebrating with a poet in a Belgrade restaurant.[14] Noticing the poet's new book of verse on the table, a patriot bought ten copies and, in a fit of bucolic confidence and to the amazement of his colleagues, Milanković vowed to study a worthy cosmic problem— the Earth's past and present climates, taking into account the tilted axis. Locked in an Austrian fortress early in the First World War, Milanković pressed on with his calculations, hardly aware of his surroundings: 'the little room seemed like the nightquarters on my trip through the universe'.[15] His results matched the known geological history of ice ages. But once again Milanković was out of luck with European politics. While a summary of his life's work was being printed in 1941, the Belgrade printing press was destroyed before the last pages could be produced.

How well did the theory of Croll and Milanković match the geological evidence? To overcome the patchy, eroded record of glacial events on land, Cambridge scientist Nick Shackleton studied sediment cores obtained from the deep ocean, where sediment had built up continuously over millions of years. In 1976 Shackleton and

his co-workers established beyond doubt that deep-sea sediments record many ice advances.[16]

The Earth has experienced five major glacial episodes—two in the Proterozoic, one in the Ordovician and Silurian (recorded in outcrops in the Sahara Desert), and one from the Devonian to Permian. The most recent (Pleistocene) Ice Age in the Northern Hemisphere lasted for more than two and a half million years, and ice sheets more than three kilometres thick flowed across northern Europe, Siberia, and North America as far south as Nebraska.[17] Valley glaciers and icecaps grew in the Himalaya, Alps, and as far south as Yosemite in California. Choked with debris from soils and uplands, the advancing ice ground down the bedrock, enlarged river valleys and cut through drainage divides. Advancing and retreating over tens of thousands of years, the ice sheets altered river landscapes far more rapidly than the drift and collision of continents ever could.

As the ice retreated a barren landscape emerged, frozen to depths of hundreds of metres. Presently the first plants found a foothold as bogs draped the glacial debris, and the Earth greened rapidly as grass, birch, and alder spread. Ice and meltwater had chiselled out the valleys of former rivers, which reoccupied and often reversed their courses. Transcontinental rivers like the Bell were destroyed and new rivers like the Mackenzie were created. Many rivers were forced to the south, and meltwater swept down the Mississippi to the Gulf of Mexico as recently as 10,000 years ago, entrenching the river and choking the channels with gravel.

When did the Ice Age end and river landscapes reappear? Newton Horace Winchell made the first estimate in 1878 using St Anthony Falls on the Mississippi at Minneapolis, where a tough limestone cap covered soft sandstone (Fig. 11.4). By the mid-1800s, the waterfall's hydraulic head was powering dozens of mills as the city became the agricultural dynamo of the northern prairies. So intense was the economic development that St Anthony Falls was soon an eyesore with logs and sawdust stuck 'in the rocky clefts intended by Nature for the joyous downward passage of crystalline waters',[18] as a visitor noted. Shafts bored into the limestone weakened the rock, and logs pounded the lip of the falls during floods. In 1869 the river broke through into a tunnel, creating a gigantic whirlpool and pushing the falls to the brink of collapse. As the news spread, barbers left

Figure 11.4 St Anthony Falls on the Mississippi River in Minneapolis.

their bristly customers and lawyers abandoned their criminal clients as the whole populace rushed to the river. A temporary dam halted the collapse, but worried city leaders knew that, with further degradation, the falls would become a rapids incapable of maintaining the mills.

How rapidly was St Anthony Falls retreating upstream? Winchell knew that, as the ice retreated, a pulse of meltwater had roared down the Minnesota River, deepening its channel and forming the ancestral St Anthony Falls at the junction with the shallower Mississippi. From there, the knickpoint represented by the falls had retreated more than 12 km up the Mississippi to Minneapolis. Using the descriptions of early travellers over several centuries (Fig. 11.5), Winchell estimated a retreat rate for the falls of about two metres a year, and he calculated that meltwater had first carved the falls about 8000 years before. He had established a date for the meltwater episode and, by extension, a minimum date for the end of the Ice Age.[19]

Some scientists were not convinced, but Winchell was later vindicated by an extinct giant beaver, the size of a bear. The creature had been rootling about under the projecting lip of St Anthony Falls near the Minnesota River when it was suddenly buried by collapsing rock. Discovered by construction workers, its bones were dated at 10,230 years old, close to Winchell's estimate. The catastrophic Minnesota River floods that first formed the waterfall are now dated to about 13,000 years ago,[20] and most northern rivers were re-established little more than 10,000 years ago.

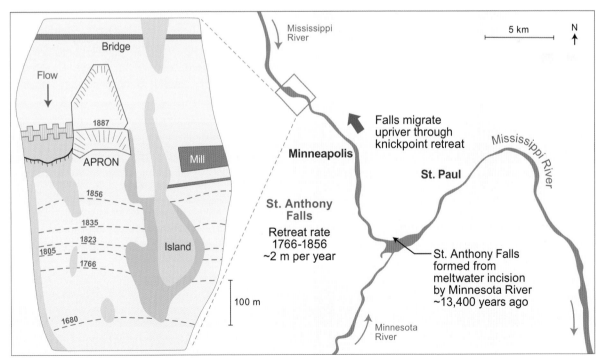

Figure 11.5 Stages in the retreat of St Anthony Falls on the Mississippi River, after Minnesota River meltwater from Glacial Lake Agassiz generated the falls following the Ice Age. After Winchell, 1878.

How the Ice Age affected the rivers of Ireland

The River Shannon drains the glacial landscape of a fifth of Ireland, surfacing from a limestone sinkhole at the small pool of Shannon Pot. Today the pot is silent, the peaty water seemingly fathomless. Ripples move out from far below—a fish, perhaps, or something more ominous. The river is named for Síonnan who came to the pool to catch the Salmon of Wisdom. But the great fish was angered and created streams that drowned the fleeing girl, forming the Shannon and casting her body into the sinkhole.[21] The Shannon now cuts through eskers of sand and gravel (Fig. 11.6) formed under the ice sheets, and the river flows to the Atlantic through a string of lakes where mayflies dance their short day of exuberant life.

Limestone forms the bedrock of half Ireland, a Carboniferous relic of a long-fled tropical sea when the country was part of Pangea. A karst network of sinkholes and caves formed as the water table lowered through geological time, and many rivers flow down into the karst in summer and are replenished from below when the water table rises during wet winters. Under Shannon Pot an underground stream wells up, travelling for kilometres past subterranean pools so still that stalactites are reflected like the towers of a drowned city. The underground streams have their own logic, with unseen drainage divides and unpredictable flow directions that often differ from those of the surface drainage above. Once, as the revenue men approached, villagers brewing illicit alcohol hurriedly threw their stills into a lake, only to find that their equipment reappeared two kilometres away in Shannon Pot.[22]

When the Cretaceous seas receded, Ireland's newly established rivers cut their courses through the slowly rising land.[23] During the Pleistocene, ice sheets deepened the upland valleys until the ice disappeared about 10,000 years ago, along with the herds of gigantic Irish elk. But the ice sheets barely eroded the lowlands, covering them instead with boulders and soil from the uplands, and many of Ireland's rivers still follow ancestral drainage courses that may date back tens of millions of years. Still visible at the surface are patches of Cretaceous river sediment that lodged in caves and sinkholes, indicating that the modern land surface is little lower than the Cretaceous level. Before the Ice Age,

Figure 11.6 Shannon River cutting through an esker formed during deglaciation at Clonmacnoise, Ireland.

the Shannon may have connected large lakes, much as it does today. It may scarcely have been a river.

Salmon falls

The little River Duff in northwestern Ireland is in spate after rain, the water frothing down the waterfalls. On the rocks in the river, young dippers are learning to feed on the larvae and nymphs of insects. The nictitating membranes over their eyes (an adaptation to underwater vision) flash in the sunlight.

In the pools below the waterfalls, salmon are leaping, spinning up from the torrent to spy-hop, or gaining a precarious hold on the ledges. Descended from the legendary Salmon of Wisdom, they returned to Irish rivers after the Ice Age and nourished the Stone Age people who colonized the barren lands. Both fish and humans left their remains as fossils. Here and there a human body is exhumed from the peat, the wrinkled skin tanned like leather and the skeleton dissolved down to jumbled bones, a rough circle of skin and hair where the skull once stood, the slack limbs curved and twisted like paper. Their deaths were violent or peaceful.

It was along the western rivers and waterfalls that the poet William Butler Yeats first heard the Irish legends that took hold in his early poetry, steeped in a world of water and salmon falls.[24] But the 1916 Easter Rising and the Civil War took their toll on Yeats even as Ireland attained long-denied nationhood. His grave in Drumcliff churchyard bears the legend: 'Cast a Cold Eye / On Life, on Death. / Horseman, pass by'.

CHAPTER 12

MEGAFLOODS AND NOAH'S ARK

Dry Waterfalls on the Columbia Basalt Plateau

The Columbia Basalt Plateau of the Pacific Northwest lies in the rainshadow of the Cascade Range. On this spring morning, the pass that crosses the Cascades is dotted with drive-through espresso bars and carwashes offering Mother's Day promotions. As the altitude increases, the Cowgirl Espresso Kiosk, replete with lightly clad cowgirls, gives way to the Chalet Espresso and the Last Chance Espresso. Exulting in the warm weather, lightly clad bikers roar by. Soon, conifer forests descend into rolling farmland where orchards grow on the rich lava soils. Meadowlarks sing from the fenceposts and magpies swoop overhead.

Harshly etched into the dark lavas through the patchwork of fields is the famous valley network of the *Scablands*. One such valley is the chasm of Grand Coulee, which drops over the basalt cliff of Dry Falls, an apparent waterfall and plunge pool that dwarfs Niagara Falls (Fig. 12.1A), with boulders the size of houses downstream. Another valley gash, Frenchman Coulee, drops abruptly over a basalt cliff where a great horned owl surveys the landscape from its nest. On the bars of the Columbia River nearby are gigantic gravel dunes, seven metres high (Fig. 12.1B).

Grand Coulee and Frenchman Coulee testify to floods of a colossal scale that crossed the lava fields and swept down the Columbia River. But there is a problem. In this dry climate, the Scabland rivers rarely flow, and the gigantic waterfalls are dry. Against prevailing views in the 1920s, J Harlen Bretz forged an understanding of a landscape formed by catastrophic floods that has profoundly influenced our understanding of flows on the Earth and Mars.

J Harlen Bretz stirs up a hornets' nest

Harley Bretz, a professor at the University of Chicago, had a gift for enthusing students about landscape.[1] He reinvented himself as J Harlen Bretz. The J stood for nothing and was not followed by a period, although typists often added one.

Bretz was a tough nut. On the first day of a two-day exam, students found ten questions written on

Figure 12.1A Dry Falls cut into basalts of the Channeled Scablands, Washington State.

Figure 12.1B Gigantic gravel dunes along the Columbia River, Washington State, formed by outburst floods. The dunes are commonly 7 m high with wavelengths of tens of metres.

the blackboard with the instruction 'Write on any five'. Arriving relaxed for the second day, they found the instruction 'Write on the other five'. But students who attended his field courses in remote areas, on foot or with pack horses, idolized him, and his legendary parties featured animal skulls with red lights in the eye-sockets.

In 1922, Bretz and his students borrowed a Model T Ford and explored the Columbia Basalt Plateau. Carved into the plateau were dramatic gorges, void of rivers or occupied by tiny creeks, with what looked like enormous dry waterfalls. In places the tough black basalt had been gouged into channels that divided and rejoined, a curious feature noted by Lewis and Clark more than a century earlier. Intervening mounds of rock stood out like scabs on skin, and there were relict 'islands' of loess that had once formed a sheet on top of the basalt but had been stripped off over much of the terrain. Bretz coined the term 'channelled scablands' for the region (Fig. 12.2). The field party visited Dry Falls and puzzled over the branching and seemingly disorganized channels. Had the ice sheets accomplished such spectacular erosion? But if water had carved these chasms, where had it come from? The next year they collected more precise data—the heights of dry waterfalls, the depths of plunge pools, the size of the gravel bars.

Bretz then wrote a paper that set the cat among the pigeons.[2] Contrary to the uniformitarian views of James Hutton and Charles Lyell, the Scablands could not be explained by the slow persistent action of rivers, he wrote. They had been forged rapidly by colossal flows of water from melting ice sheets, of an intensity hitherto unsuspected. The flood waters had reached heights of 200 m above the modern level of the Columbia, biting deeply into the basalt. A debacle had swept the plateau, and Bretz later estimated a flood discharge of two million cubic metres a second. But he confessed to 'a feeling of amazement that such huge streams could take origin from such small marginal tracts of an ice sheet'.[3]

For a while all was quiet. Then in 1927 Bretz was invited to present his results at the elite Cosmos Club in Washington, DC,[4] of which John Wesley Powell had once been president. Before the most influential American geoscientists of the day, Bretz described his field sites, which he felt should speak for themselves. To explain the source of the water, he drew on the experience of a Swedish student at Chicago, Hakon Wadell, who had visited the Vatnajökull Icecap of Iceland and barely survived an intense storm that blew away two of their ponies.[5] The field party found that the Grímsvötn volcano had erupted through the icecap, creating a vent filled with water that periodically released catastrophic meltwater floods. Perhaps, Bretz

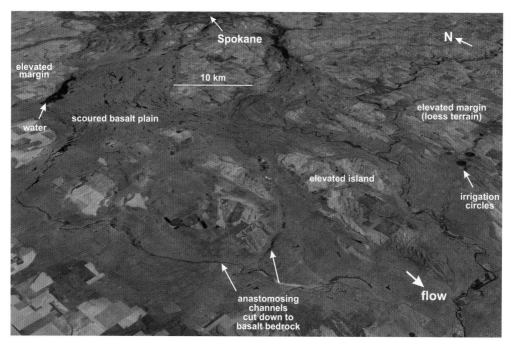

Figure 12.2
Part of the Channeled Scablands of Washington State, incised through a loess blanket into bedrock of the Columbia Basalt Plateau. Maps data: Google, Landsat / Copernicus.

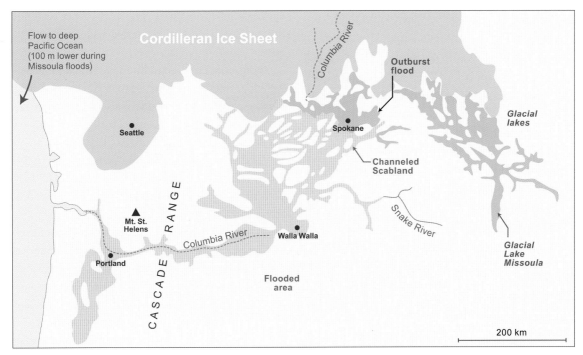

Figure 12.3 Channeled Scablands and glacial lakes bordering the Cordilleran Ice Sheet. The outburst floods were caused by collapse of an ice dam from a lobe of the Cordilleran Ice Sheet, which held back Glacial Lake Missoula. After Waitt et al., 2019.

reasoned, active volcanoes below the ice sheets had caused the Scabland floods.

Then the scientists laid into him, as they had always intended to do. They advanced every other possible explanation—channels formed by collapsed lava tubes, erosion from a Columbia River diverted by glaciers, prolonged wearing down by rivers no different than today. And the water—impossible to generate so much in a short time and impossible to carve into solid basalt so quickly. Some were polite, others were hostile and condescending. Few had visited the area, but it didn't seem to matter. An unwarranted catastrophe had entered the comforting landscape of uniformitarianism and had been treated with the severity that it deserved.

For a long time afterwards Bretz was deeply depressed. He sat at home in the evenings smoking his pipe, but his feet and hands shook. His wife was heard to say, 'If you know you are right, Harley, then nothing anybody else says can hurt you.'[6] Finally Bretz regrouped. He mailed off letters and found support from scientists who also had defied the establishment. He moved on to a study of caves. His ideas also seemed to go underground.

Then in 1942 survey geologist Joseph Pardee published a study that cast new light on the controversy.[7] Pardee had been present at the Cosmos Club confrontation, and he had whispered to a neighbour that he knew the source of Bretz's flood. Glacial Lake Missoula had once filled a large valley east of the Scablands, blocked by a tongue of ice (Fig. 12.3). In the valley, Pardee found gouged rock and gigantic gravel dunes, and he suggested that Glacial Lake Missoula's ice dam had failed, releasing a catastrophic flood. He estimated that the lake had been 600 m deep with a volume of 2000 cubic kilometres—the size of a small Great Lake.[8]

Gradually the tide turned in Bretz's favour. Catastrophic floods were now understood as natural mechanisms, sculpting courses that no repeated drip of water on rock could have accomplished. In 1965, an excursion visited the area. Bretz did not attend, but he asked the trip leader to leave eight unoccupied seats for sceptics, most of whom had been present at the 1927 meeting but had gone unrepentant to meet their Maker. Presently a telegram reached him from the excursion members with the words 'We are all now catastrophists.'[9]

Megafloods in North America and Siberia

How large were the floods that swept out of Glacial Lake Missoula and over Dry Falls?[10] More recent work suggests that the lake may have been deep enough to float the ice dam, releasing water below, or the lake may have burst through collapsing ice. In any event a wall of water hundreds of metres high would have roared down the valley at a speed of more than 100 km an hour. For perhaps a few days, as much as 20 million cubic metres roared across the Scablands every second, far more than Bretz had estimated and equivalent, while it lasted, to more than fifty Amazons at peak flood and some twenty times larger than all the rivers on Earth combined. Offshore from the Columbia river mouth, the floods travelled for more than 1000 km under the sea, carving gorges through the ocean sediment and laying down nearly 200 m of sand in deep water. Bretz had envisaged just a few floods, but cliffs along the Columbia contain thick stacks of sand layers, each of which was laid down by a colossal flood (some 80 in total) as sand was swept

upstream into tributaries (Fig. 12.4). After each flood, the ice lobe would have advanced again and rebuilt the dam.

Indigenous cultural traditions across North America include accounts of gigantic beavers and the damming of lakes, especially in areas where glacial lakes were common. They may be oral records of meltwater pulses.[11] The Scablands floods probably took place between 17,000 and 13,000 years ago as the Ice Age waned, and it is unclear whether humans were present in the area. J Harlen Bretz may have been the first person to 'see' the Scabland floods.

Bretz's flood was far from the isolated occurrence that he had supposed.[12] The Laurentide Ice Sheet and the other ice sheets of North America with which it merged (Fig. 12.5) melted (deglaciated) from about 22,000 to 7000 years ago, generating large glacial lakes (superlakes) in depressions across the northern part of the continent (Fig. 9.2). Dammed by ice and glacial sediment, the lakes frequently burst out in gigantic floods that swept river valleys, sometimes powerful enough to flow uphill—a rare exception to Paul Potter's

Figure 12.4 Sheets of sand and silt near Walla Walla, Washington State, each formed when an outburst flood through the Scablands backed up flow in the Walla Walla River.

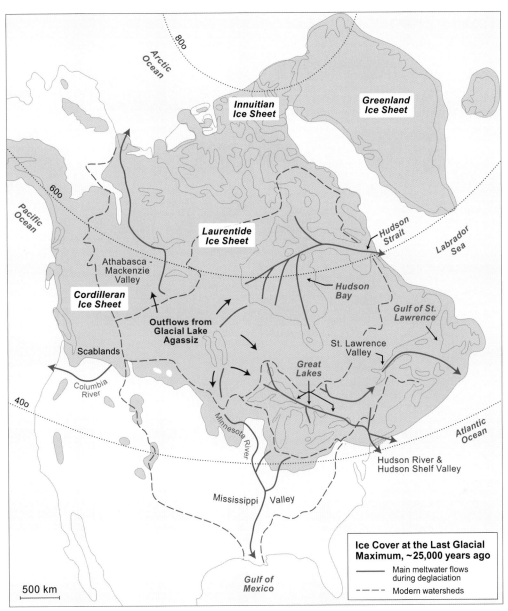

Figure 12.5 North American ice cover during the Last Glacial Maximum, and drainage pathways during deglaciation, including runoff and outbursts from Glacial Lake Agassiz and other glacial lakes. After Lewis and Teller, 2007.

dictum (Chapter 4). The modern Great Lakes are a remnant of these superlakes, formed where a succession of glacial episodes carved deeply into river valleys. Among the largest lake was Glacial Lake Agassiz, more than 160,000 cubic kilometres in volume—some thirteen times the volume of Lake Superior. The lake drained catastrophically across Hudson Bay, into the Great Lakes, and out to the Arctic Ocean, sending meltwater roaring down the Mississippi, St. Lawrence, Hudson and Mackenzie river systems. A dramatic outburst from Glacial Lake Agassiz more than 8000 years ago sent an armada of icebergs through Hudson

Strait and far south into the Atlantic Ocean, radically altering the Earth's climate.

Across Asia, the deglaciation of the Siberian ice sheets also generated catastrophic floods. Well documented deglaciation events nearly 100,000 years ago generated extensive lakes south of the ice front, flooding the Yenisei and Ob river valleys east of the Urals and finding spillways through to the Baltic Sea and through the Aral Sea to the Caspian and Black seas (Fig. 12.6). Collapsing ice dams generated flood waves hundreds of metres deep, forming gravel ripples with crests twice the height of a house.

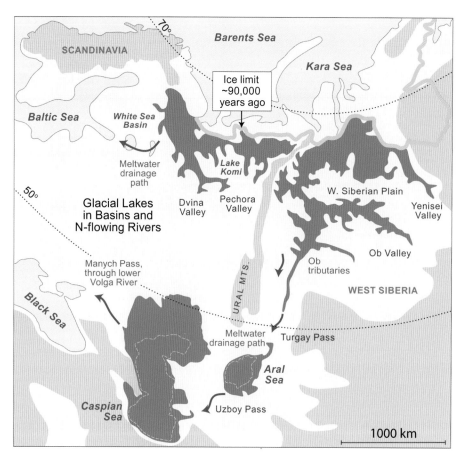

Figure 12.6 Glacial lakes and outflow paths from the Siberian Ice Sheet to the Black Sea and Baltic Sea, about 90,000 years ago. After Mangerud et al., 2001.

Megafloods in Iceland

The road around southern Iceland crosses a *sandur* below the Vatnajökull Icecap (Fig. 12.7), a braided-river plain of black volcanic sand and gravel tens of kilometres wide (Fig. 12.8). In 1996, Grímsvötn erupted under the icecap and, more than a month later, a meltwater flood roared down the sandur, transporting ice blocks that weighed up to 2000 tons. Flowing at 50,000 cubic metres a second, this relatively small megaflood was briefly the second largest river on Earth,[13] tearing out the road and sweeping bridge girders downflow. The destruction of infrastructure was expected: no engineer with a realistic budget could design for such a flow, and no permanent road crossed the sandur until the 1970s.

The glacial river of Jökulsá á Fjöllum flows north from Vatnajökull and thunders in a vortex of spray over Dettifoss, Europe's most powerful waterfall (Fig. 12.9A). Downstream and now disconnected from the river is the curved cliff of Ásbyrgi, cut through the lava flows with a deep plunge pool (Fig. 12.9B).[14] Like Dry Falls, no river runs over the cliff, but there is familiar evidence that enormous floods cut Ásbyrgi: a litter of gigantic boulders, a basalt scabland, and enormous gravel ripples. Modern flows on the river seldom reach 1000 cubic metres a second, but Ásbyrgi was sculpted by floods that are estimated at 700,000 cubic metres a second.

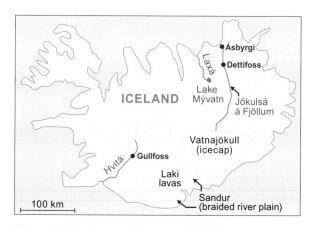

Figure 12.7 Rivers and ice-caps of Iceland.

Figure 12.8 Skeiðarársandur (outwash plain) of dark volcanic gravel and sand, southern Iceland. The outwash plain formed by meltwater outbursts from eruptions of the Grímsvötn volcanic edifice below the Vatnajökull Icecap, seen in the background.

Figure 12.9A Dettifoss waterfall on the Jökulsá á Fjöllum River, Iceland.

Figure 12.9B Dry waterfall at Ásbyrgi, Iceland, cut by meltwater outbursts estimated at 700,000 cubic metres a second, caused by volcanic eruptions under the Vatnajökull Icecap.

Noah makes a comeback

Noah, of course, began it all. Deemed the only man of integrity on Earth, he became an unwilling shipbuilder, animal collector, and sailor. In the biblical account, Noah's ark drifted on the Great Flood or Deluge, later allowing people and animals to return to the land under the sign of the rainbow. Fossils were widely attributed to the Deluge, although Leonardo da Vinci argued that fossil shell accumulations resembled those found on any ordinary beach. His conclusion depended on judg-ment and reason which, in the words of Shakespeare in *Twelfth Night*, 'have been grand-jurymen since before Noah was a sailor'.

But the Deluge lived on in the minds of early scientists.[15] In the ruins of Nineveh, workers falling through a ceiling discovered a royal library from the seventh century BCE. In 1872 while working with thousands of fragments of clay tablets, a young assistant George Smith was amazed to read in the Akkadian script of a ship resting on the mountains, followed by the sending forth of a dove. Rumour has it that Smith, startled to have confirmed the biblical flood, rushed out and began to tear off his clothes, to the surprise of passersby. Smith subsequently found more fragments as part of the *Epic of Gilgamesh* in which the Great Flood is released because, with the burgeoning human population, 'the uproar of mankind is intolerable and sleep is no longer possible by reason of the babel'.[16]

In 1998, geologists Bill Ryan and Walter Pitman published *Noah's Flood* after working with sediment cores from the Black Sea. They inferred that melting ice raised global sea level and forced the Mediterranean over a gigantic waterfall into the Black Sea. The megaflood would have raised the level of the Black Sea by 15 cm a day, they suggested, and drowned the borderlands within two years. Traumatized farmers, among them Noah, would have escaped to the Fertile Crescent and elsewhere, carrying with them their haunting memories of the Great Flood. However, the evidence for such a megaflood has been strongly challenged.[17]

But there is another possibility for Noah. As the Siberian ice sheets retreated at the end of the Ice Age, meltwater lakes across Eurasia dwarfed even Glacial Lake Agassiz. As the lakes emptied out, megafloods erupted through glacial spillways and river valleys into the Aral, Caspian, and Black seas.[18] Such catastrophic floods may have entered permanently into the human consciousness.

CHAPTER 13

RIVERS DROWNED BY THE SEA

Drowned rivers of Arctic Canada

At Pressure Point on the northwest tip of Somerset Island, sea ice drifting through the Northwest Passage has forced up ridges of jumbled ice blocks. Even as spring advances in the Canadian Arctic Islands, this remains a forbidding place, and Peel Sound to the west is still covered with pack-ice. Where the frozen ocean and snow-covered land meet is anyone's guess (Fig. 13.1A).

The De Havilland Otter has put down on the gravel beach and headed for home, its red tailplane a receding speck. The only human presence is three tents, their guy ropes wrapped around boulders, placed in shallow pits excavated in the permafrost. But circles of stones and bone show where resourceful Inuit hunters pitched their tents long before.

The short Arctic summer unfolds, and Peel Sound slowly clears, the ice masses booming as they run the tide under the midnight sun, the only sound in a silent land. Narwhal thrust their tusks up through leads in the ice, and belugas move into the bays. Other animals appear—a polar bear tracking seals, a herd of musk ox, an arctic fox that eats the webbing of the snowshoes. A group of ermines is so unafraid that the animals leap onto a shoulder and rummage in a pocket. Two months from now, the sun will dip below the horizon and Peel Sound will freeze over again.

In 1846, *Erebus* and *Terror* under the command of Sir John Franklin sailed down an ice-free Peel Sound and were trapped in the ice for the following two years. With little alternative, the crew headed south on foot, dying from starvation and disease, with lead poisoning from soldered food cans as a contributory factor. Over the next decade, shipboard and sledge parties searched for Franklin, gradually mapping the islands and sounds.

Lieutenant George Mecham led one of the search parties. His men scoured the northwestern archipelago

Figure 13.1A Peel Sound (ice-covered at left) at Pressure Point, Somerset Island. The sound was originally part of a river system in the Canadian Arctic, later deepened by ice and drowned by post-glacial sea-level rise.

Figure 13.1B Wood material in Pliocene river sediments of the Beaufort Formation at Ballast Brook on Banks Island, Arctic Canada. The Beaufort Formation formed a wedge of sediment that built into the Arctic Ocean, until lowered sea-level at the onset of the Ice Age caused entrenchment of the river systems. Courtesy of John Gosse.

through the winter of 1853, dragging a sledge for more than 2000 km across the ice and hammered by snowstorms. When the brief summer thaw set in, melting permafrost and torrential rivers added to their misery. Then, to his amazement, Mecham found wood projecting from gravel beds in the cliffs of Prince Patrick Island (Fig. 13.1B), including a trunk 10 m long. He kept a fragment on the sledge. 'I could not but fancy (strange as it seemed),' Mecham commented about the forest, 'that it had grown in the country'.[1]

Mecham had found fossils of the Beaufort Formation of probable Pliocene age (Fig. 13.2), laid down by braided rivers that flowed northwest to the Arctic Ocean.[2] The river channels were choked with wood, some of which had been gnawed by beavers and formed part of the earliest known beaver dams. With the wood were fossils of bears, frogs, and giant camels. As Mecham rightly surmised, the trees had grown nearby but under warmer conditions than those of the modern

Arctic, flourishing in the long hours of daylight at high latitude. But some of the associated river cobbles bore tell-tale glacial striations from ice buildup and meltwater in the early days of the Ice Age.

How had the sounds formed between the islands? The question was largely resolved by geologist Bernie Pelletier, a tough campaigner who had fought through Europe in the Second World War and once, after a long field season in the Rockies, had ridden his horse into a bar and bought it a drink.[3] He survived a helicopter landing that shattered the machine: 'It's rather drafty in here', said Pelletier from the wreckage. Pelletier determined that the sounds between the islands were the drowned valleys of an ancestral river system, with a drainage divide close to Somerset Island.[4] A river had flowed north between Canada and Greenland, and the ancestral Coppermine and Back rivers of the mainland had once joined the system. As sea level fell at the onset of the Ice Age, glaciers had occupied and deepened the

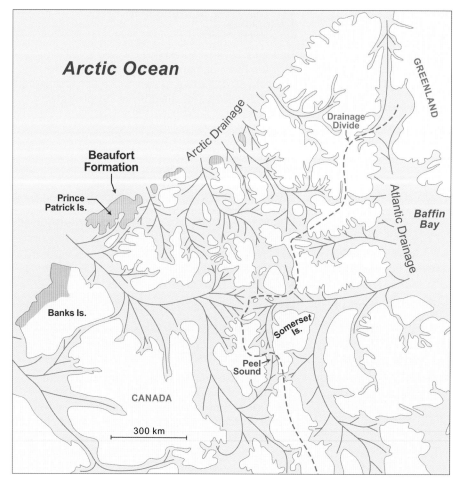

Figure 13.2 Drainage system of the Canadian Arctic early in the Ice Age. As ice began to build up and sea-level lowered, late Miocene to Pliocene river sediments of the Beaufort Formation were cut through by the rivers that had formed them. Their courses were later deepened by glacial erosion and then drowned as the Laurentide Ice Sheet melted within the past 20,000 years. After Fortier and Morley, 1956, and Bornhold et al., 1976.

Figure 13.3 Temple of Serapis at Pozzuoli in the Bay of Naples, Italy, which shows relative uplift and subsidence linked to ground deformation in this volcanically active region. The dark mid-part of the pillars, about 3 m in vertical extent, was bored by marine clams when the temple was drowned (the un-bored lower part was covered with sediment).

river valleys, slicing through the Beaufort Formation. When the Northern Hemisphere ice began to melt some 22,000 years ago, the rising ocean drowned the valleys to form the modern sounds. Similar drowned, glacially deepened river valleys are widespread as fjords in North America and Scandinavia.

People had long known that sea level rose and fell.[5] Underwater ownership had vexed communities for centuries, and some Scottish jurisdictions allowed land to be claimed below the sea as far out as a horseman could ride and then throw a spear. The Vikings created special land laws along the Baltic Sea, where new land was visibly emerging through glacio-isostasy as the weight of ice was removed. Swedish astronomer Anders Celsius, inventor of the temperature scale, interviewed hunters about water levels on a rock where seals rested, scratching recording marks on the stone, and Charles

Lyell added new scratches in 1834. As the frontispiece of *Principles of Geology*, Lyell featured a Roman temple at Pozzuoli near Naples in Italy, where marine clams had bored holes on the pillars when the volcanic area subsided below the sea (Fig. 13.3).

As the ice sheets melted, large volumes of water re-entered the oceans, as Charles Maclaren had surmised, and the water expanded as the ocean temperature increased. Many rivers on coastal shelves were drowned in little more than a few thousand years. By about 7000 years ago the ice sheets were largely gone, sea-level rise slowed, and the rivers readvanced to construct new deltas.[6]

Three rivers stand out for the sheer magnitude of the drowning. The *Molengraaff River* once drained much of offshore Indonesia; the *Eridanos River* lies under the Baltic Sea; and the *Channel River* formerly ran down the English Channel to the Atlantic.

Molengraaff River of SE Asia

The Molengraaff River once crossed the Sunda Shelf, almost two million square kilometres in area, which linked Sumatra, Borneo, and Java with mainland Southeast Asia when ice sheets were extensive and sea level was low (Fig. 13.4). Dutch geologist Gustaaf Molengraaff was well known for describing the Cullinan diamond of South Africa, the largest ever discovered.[7] While mapping the ocean floor off Indonesia, Molengraaff discovered that the rivers of Sumatra and Borneo could be followed under the Sunda Shelf as valleys tens of kilometres wide, maintained as hollows by strong ocean currents. Some were rich in the mineral cassiterite, eroded from onshore tin deposits. In a 1921 publication, he inferred that they had once been the headwaters of a drowned river tract that included the lower reaches of the Mekong and Chao Phraya rivers. The Molengraaff River, named for its discoverer, ran through a vast tropical forest that covered the Sunda Shelf.[8] When the ice sheets melted, sea level rose at a metre a century, driving back the river by more than 200 m a year until it was reduced to disconnected island streams.

Victorian scientist Alfred Wallace contributed to this insight from his study of the island organisms in the vicinity of the Sunda Shelf.[9] He mapped a boundary (later termed Wallace's Line) through the islands, separating regions with largely Asian or Australian faunas. It was tough work: he mistook a poisonous snake for his handkerchief one night, and marauding ants tore his insect collections from gummed cards. While laid low with malaria, he formulated a theory of natural selection that he later presented with Charles Darwin.

The drowning of the coasts has remained in the oral traditions of Aboriginal communities around Australia.[10] Some traditions speak of the sea covering the coast after a giant kangaroo cut a trench with a magic bone, or of a forest that burned with such intensity that the land split and the sea came in. A thirsty man jabbed a hole in a skin waterbag, drowning along with the land. Considering the local water depth and the rate of sea-level rise, the traditions may record events as much as 13,000 years ago. Knowledge of such dramatic changes could have been passed down among people who were embedded in their local landscapes and listened to the land.

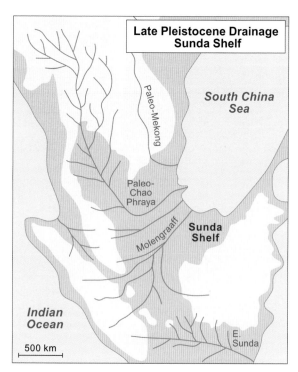

Figure 13.4 Map of the Molengraaff River and other drainage systems drowned by post-glacial sea-level rise across the Sunda Shelf. After Alqahtani et al., 2017.

Eridanos River of the Baltic

Every hour of the day and night, a trumpeter plays an incomplete burst of notes from the top of the church tower in Kraków's main square. The notes honour the trumpeter who in 1241 warned the populace of a Mongol attack on the Polish city until he was suddenly cut off by an arrow through the throat. Below the square lies the excavated rubble of the medieval city, which includes pieces of amber from necklaces and bracelets that glow with a strange brilliancy in the subterranean gloom.

Amber caught the eye of Stone Age people across Europe and North Africa, and it was believed to heal diseases and protect warriors.[11] But what was amber? Greek mythology was rife with suggestions. Was it the solidified tears of sisters on the banks of the legendary Eridanos River for a brother killed by lightning? Or was it the solidified urine of a lynx? Writing in the first century CE, Pliny the Elder poured scorn on such mythical explanations.[12] In his view amber was the resin of a pine tree, hardened and thrown up along the shores of a northern ocean. Amber smelled like pine when burnt and contained insects entombed in the once-liquid droplets. Pliny did not identify the Eridanos River, but the name may come from the Radanus river, a Baltic tributary of the Vistula.

By far the world's largest amber accumulation lies along the Baltic Sea coast. There, under a tropical climate in the Eocene some 40 million years ago, the copious resin of coniferous trees hardened into amber through geological time. Baltic amber has yielded some 250,000 fossils and more than 3000 species of animals. Centipedes, bees, scorpions, spider webs, and geckos: all were entombed in the sticky resin. Mining and dredging produced 500,000 tons of raw amber annually up to the start of the First World War, and Peter the Great of Russia was given an entire room filled with amber sculptures and mosaics. Of incalculable value, the amber room was seized by German forces during the Second World War and crated up as the Russians advanced, vanishing without trace.

From the Miocene onwards the *Eridanos River* had drained much of northwest Europe, following the line of the modern Baltic Sea. On its way to the North Sea, the river picked up the ancestral Vistula, Oder, Elbe, Weser, and Rhine-Meuse rivers from the rising Alps and Carpathians (Fig. 7.3). At its peak, the Eridanos drainage system was similar in size to the modern Ganga or Orinoco, transporting large amounts of Baltic amber to western Europe.[13]

About a million years ago, ice sheets covered much of the river's Scandinavian headwaters and deepened the valleys. When the ice sheets melted, the Baltic Sea filled the area. The former Eridanos tributaries became independent rivers, and after a lifespan of more than 30 million years, the Eridanos River was gone.

The Channel River of Western Europe

The German High Command employed some 250 geologists by the end of the First World War, but the British War Office didn't seem to get it. By mid-1916, only geologists Bill King and T.W. Edgeworth David were deployed on the Western Front.[14] There, they played an unexpected role in establishing the history of the Channel River that once ran down the English Channel through Dover Strait. Yorkshireman Bill King was awarded the OBE for his wartime work and became a kindly university professor.[15] Antarctic veteran

Edgeworth David was a white-haired grandfather who founded the Australian Mining Corps and enlisted with his troops, constructing dugouts and tunnels. Said to be the politest man who ever came out of Australia, David would address his batman in terms such as 'Johnson, I have left my Sam Browne in my billet. I wonder if you would mind stepping over for it, that is if you have time.' His commanding officer remembered his first meeting with David, a tall man standing strictly to attention.

I said 'You have done a lot of exploring.' 'Yes, sir.' 'Do you like it here?' 'Yes, sir.' Then I said to him, 'Look here, David, for Christ's sake sit down and stop embarrassing me by standing to attention. You have done 20 times more in your life than I shall ever do.'[16]

The Somme trenches snaked across uplands of porous Cretaceous Chalk with dry valleys but little surface drainage. The valleys formed by flow over impermeable permafrost during and after the Ice Age, drying up as the permafrost melted and the water table lowered. King and David had to drill to obtain water for the troops and draft animals.[17] Using French maps, King found that the Chalk contained folds with arch-like *anticlines* and trough-like *synclines* that influenced the height of the water table, and he set up a framework for linking the Somme ridge on the *Artois Anticline* (Fig. 13.5) with the *Weald Anticline* of England across the Channel. In Belgium, the trenches cut through the

Figure 13.5 Somme Battlefield at Thiepval, France. The German defensive line during the First World War followed the ridge crest of Cretaceous Chalk on the Artois Anticline, above the Ancre River (out of sight in the middle distance).

Eocene Ypres Clay, where shallow wells flowed too slowly to supply large troop buildups. King discovered that the aquifers were under sufficient pressure to force the water to the surface—an artesian system, named for Artois—and he designed sieves to trap the sand that impeded water flow. On the Passchendaele ridge, the German Army had water problems of a different kind because the Eocene Paniselian Formation became quicksand during construction or under shell fire. During the last days of the war, King's men drilled wells under enemy fire to supply 300,000 men and 100,000 horses advancing across a waterless plain east of the Somme trenches. But not all geological work had such practical motives. While excavating a dugout, crews found the remains of a mammoth, and David brought back a tooth and chipped flints. Everyone was talking about the mammoth and, for a moment, the war seemed far away.

How and when had the Channel cut through the Weald–Artois ridge? A geological graduate of the First World War, Dudley Stamp obtained a geology degree on leave from the trenches, and he carried out fieldwork between military operations.[18] He had a keen eye for the land, and his obituary noted that he was now exploring an uncharted country for which he was neither unprepared nor a stranger. Stamp followed up King's work by correlating the strata on either side of the Dover Strait. The Ypres Clay matched the London Clay, excellent tunnelling rock for the London Underground, and the Paniselian Formation matched the Eocene Bagshot Beds on Hampstead Heath north of the city. Dover Strait could have formed only after these strata were laid down and was no older than the Eocene. Might river capture have cut through the Weald–Artois ridge, widening into Dover Strait? But the ridge was narrow, and any rivers would have been small.

Called up to give geological advice early in the Second World War, Bill King was caught up in the retreat to Dunkirk in 1940, receiving the Military Cross for guiding a truck convoy of high explosives through roads with which he was familiar.[19] It was partly on his advice that Normandy was chosen for the invasion, because it furnished ground suitable for airfields. Working on Operation Pluto, a wartime oil pipeline to France, King found surprisingly little information about the geology of the Channel. After the war, he and his colleagues made an undersea geological map using the new method of seismic refraction to image the underwater strata, and deploying a bomb-like device that drove a steel tube into the seafloor to obtain samples. They found that much of the Channel followed old faults—lines of weakness that rivers had exploited since the Cretaceous sea receded.[20] Across the uplands of France and England, the rivers cut valleys into the Chalk and tropical soils mantled the adjoining land, surviving mainly as the residual sarsen stones used by the Neolithic builders of Stonehenge.[21] But the Weald–Artois ridge crossed the sea unbroken by faults and had no early river history. Dover Strait was an anomaly.

By the 1980s, bathymetric surveys showed that the English Channel was underlain by the course of a young river—the *Channel River*, 75 km wide, joined by drowned tributaries from France and England (Fig. 13.6).[22] The Channel River comprised the narrow Lobourg Channel through Dover Strait, Northern and Median palaeovalleys, and the deep, narrow chasm of Hurd Deep, more than 150 km long north of Cherbourg. The Northern Palaeovalley was cut into bedrock as a 'box cut' with a flat floor, streamlined walls, and bordering benches. Within the palaeovalley, a suite of channels anastomosed around islands of Chalk bedrock.

As new information emerged, a remarkable account took shape. About 450,000 years ago, advancing ice sheets pushed the Thames south to its modern course through London, and a large meltwater lake built up in the southern North Sea, fed by the British and Scandinavian ice sheets, the Rhine, and the Thames. The lake was held in place by the narrow isthmus of the Weald–Artois ridge, with a Chalk escarpment on the northern side (the North Downs in England). As the lake deepened, the water cascaded over the Chalk escarpment as waterfalls and cataracts with downflow plunge pools. The plunge pools, discovered in the 1960s and '70s during investigations for constructing the Channel Tunnel, are termed the *Fosses Dangeard*— amphitheatres as deep as 140 m in a narrow belt across the Dover Strait, below knickpoints in the topography. Eroded in numerous places, the ridge collapsed and megafloods with a flow rate of a million cubic metres a second poured from the lake and roared through Dover Strait, sculpting streamlined bedrock islands like those of the Scablands. The megafloods plunged over the continental slope of the Bay of Biscay, transporting a large volume of sediment to the deep Atlantic.

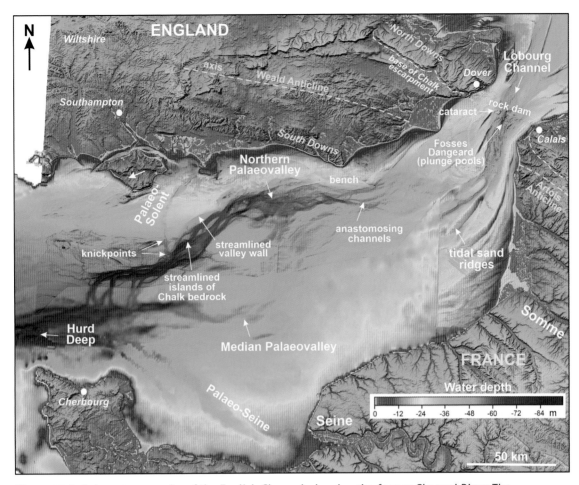

Figure 13.6 Subsea topography of the English Channel, showing the former Channel River. The river formed largely by megafloods after the Weald–Artois ridge was breached in the Dover Strait by overspill from a glacial lake in the southern North Sea. After Gupta et al., 2007.

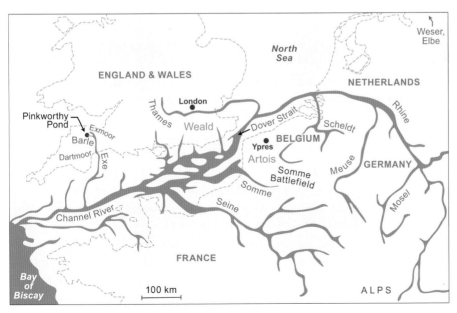

Figure 13.7 Channel River, which united many of the rivers of Western Europe before its lower course was drowned by post-glacial sea-level rise, leaving its tributaries as disconnected modern rivers. After Gibbard, 2007, and Bridgland, 2003.

Once the ridge was breached and the lake drained, the Channel River captured most of the adjoining rivers, including the Rhine, Meuse, and Scheldt and possibly even the Elbe, Weser, Oder, and Vistula (Fig. 13.7), diverted by northern ice. At its peak, the river system flowed for 2000 km from the Rhine's Alpine headwaters and carried half the drainage of western Europe to the Atlantic, with a catchment area the size of the Ganga. Many of the tributary valleys contain staircase terraces, each terrace matching a former level of the Channel River downstream.[23] Especially famous are the Somme terraces, which yield stone axes at St. Acheul near Amiens (source of the Acheulian handaxe style) and were studied among others by scientist, politician, and banker John Lubbock, first Baron Avebury.[24] Lubbock drew his title from a Stone Age site and introduced the terms Paleolithic and Neolithic (Old and New Stone Age). When his father brought important news, the youthful Lubbock was disappointed to learn that someone called Charles Darwin had moved to the neighbourhood, not the new pony he had expected. They later became close friends.

As the ice sheets melted some 20,000 years ago, rising sea level drowned the Channel River. The tributaries were reduced to disconnected headwater streams, and the Palaeo-Solent ceased to exist when Southampton Harbour was drowned. 'Island Britain' was separated from Europe, influencing travel and politics from the Stone Age to Brexit.

Chalkland rivers

As a teenager, I spent a year in the heart of the Weald Anticline in Sussex, close to the English Channel. Memories of the Second World War were still vivid. I worked with a man who had entered the local town square and come face to face with a German V-1 rocket, caught in the tramwires a few metres off the ground. Nightingales poured out their songs by day and night, descendants of the birds that were recorded singing above the roar of bombers returning from the Ruhr. Weekends found me cycling through the Cretaceous chalklands with a geological map, recording observations in a Pocket Jotter with a more assured style than my childhood diary entry for 23rd November 1963: 'A sunny morning but a cloudy afternoon. Our

hamster, Titian, has died. President Kenneddy (sic) has been assassinated.'

What did a teenager hope to achieve with the jotter, and what was in his mind as he cycled wearily home? I cannot find that I was laying a foundation for a later career. I was simply enthralled with the origins of a landscape that long predated human history. To observe and record was to understand, or so I hoped.

Long before the work of King and David, chalkland naturalists pondered the history of the landscape and the streams that were tributaries to the Channel River. Pre-eminent was Richard Jefferies, who grew up on a Wiltshire farm near the prehistoric route of the Ridgeway, among ancient tumuli 'alive with the dead'.[25] Employed as a journalist, he travelled widely in the chalklands and read Darwin and Lyell. After moving to suburban London, Jefferies began to write about what he knew best—the natural world of the chalklands—and he came to public attention with *The Gamekeeper at Home*, an authentic portrait of a local keeper. He wrote about the thrushes breaking snail shells on the sarsen stones at the family farm, and about the chalkland valleys, dry in summer but flowing as *winter bournes* to the Thames as the groundwater was replenished and springs appeared (Fig. 13.8). Jefferies later moved to a cottage on the Sussex coast, where he died of tuberculosis in 1887 at the age of 38.

Most remarkable of all was his mystical vision at an Iron Age hillfort, keenly aware of the silent company of prehistoric dwellers and ancient creatures:

> I was not more than eighteen when an inner and esoteric meaning began to come to me from all the visible universe … in the grass fields, under the trees, on the hill-tops, at sunrise, and in the night. There was a deeper meaning everywhere. The sun burned with it, the broad front of morning beamed with it; a deep feeling entered me while gazing at the sky in the azure noon, and in the star-lit evening. I was sensitive to all things, to the earth under, and the star-hollow round about; to the least blade of grass, to the largest oak … I saw back through space to the old time of tree-ferns, of the lizard flying through the air, the lizard-dragon wallowing in sea foam, the mountainous creatures, twice-elephantine, feeding on land; all the crooked sequence of life … The immense time

Figure 13.8 Dry valleys at the Devil's Punchbowl on the Ridgeway, UK. The valleys formed by runoff over permafrost during and after the Ice Age, after which the water table lowered in the porous Cretaceous Chalk.

lifted me like a wave rolling under a boat … Like a shuttle the mind shot to and from the past and the present, in an instant.[26]

In the 1930s, my grandfather Robert Gibling, still suffering from the effects of poison gas and never referring to his wartime experience, used *The Gamekeeper at Home* for his school classes. The gamekeeper's cottage is still as Jefferies described it,[27] but the man-traps are gone, outlawed even in Jefferies' day. The gamekeeper would not have approved of the red kites nesting nearby, but the present owner's sculpture of a pheasant trampling on shotgun cartridges might not have displeased him.[28]

Jefferies would have loved the thought that his cottage once lay on the bank of a great, uniting Channel River, with the Chalkland bournes of the Ridgeway as its tributaries. He would have felt a kinship with the Paleolithic wanderers who left their stone axes by the Somme, as alive to him as though they had lived only moments before. I place a flint by Jefferies' grave at Worthing, removing my shoes and socks in honour of a man who wished to bring up 'children whose naked feet are not afraid of the dew'.[29]

I write the final words of this chapter by the Rhine in the German city of Mainz, where barges are beating upstream in the last of the daylight. In my imagination, I see an international river that gathers the water of Europe but, for now at least, has no existence. I see the Rhine rising in the Alps and collecting the Mosel, Meuse, and Scheldt en route to its confluence with the Thames. The river threads the Lobourg Channel through Dover Strait before gathering the Somme and the Seine and thundering through Hurd Deep to the open Atlantic.

PART 4: HUMANS AND RIVERS

CHAPTER 14

FROM STONE AGE STREAMS TO RIVER CIVILIZATIONS

Hominins explore their rivers

On the plains of East Africa, an ape-like creature walks slowly along a low ridge by a lake and gazes apprehensively at the smoking volcano of Ngorongoro, which had erupted violently the year before, filling the lake with ash and lava. Dusk is descending over the Serengeti Plains with their vast herds and seasonal rivers.[1]

The creature scratches and urinates. Then it picks up a piece of pink quartzite carried down by an upland stream. Cobbles are scattered along the ridge—dark lava that generations of the creatures have carried up from the channels. There are corestones with facets where flakes were knapped off, as well as pebbles pitted from their use as tools. Nearby lie the bones of giraffe and hippo, brought by the big cats that hunt relentlessly.

Carnivore dung, decaying in the equatorial heat, litters the ground. Picking up a quartzite flake with deft hands, the creature neatly fillets a bone with a lens of flesh still attached.

A cat growls among the distant palm trees. The creature shivers and judges the distance to the nearest tree. There are crocodiles in Palaeolake Olduvai, where freshwater springs are an oasis in the harsh landscape. Seasonal rains sometimes flood the ridge: then the trees are a real refuge.

Homo habilis or 'handy man' roamed Olduvai Gorge in Tanzania (Fig. 14.1) nearly two million years ago. The creatures were omnivorous, for *Homo*, then as now, is nothing if not opportunistic. As Tagore pointed out, the

Figure 14.1 Olduvai Gorge in Tanzania, prominent site of hominin evolution. Dark volcanic rock about 2 million years old forms the base of the gorge in the middle distance, and the overlying strata are composed of volcanic detritus reworked by lakes and rivers.

creature's upright stance denotes an inherent insubordination.[2]

Homo habilis was not the first of our kind. Hominins were present in Chad more than seven million years ago, and some of the earliest known stone tools were made in Kenya more than three million years ago.[3] *Australopithecus* later made tools from Ethiopian river cobbles with a childlike eye for beauty, selecting translucent quartz and volcanic rocks with their coloured crystals. Crouching among stranded logs and alert for crocodiles, they cut the meat from animal bones and eviscerated carcasses. Some 200,000 or more years ago, *Homo sapiens* appeared.[4]

The Olduvai toolmakers shared with us their DIY skills and perhaps a language ability to rival the chattering and twittering classes. Like monkeys at the zoo, they would have seemed tantalizingly human and disturbingly vulgar. The Divine Being invented humans because he was disappointed in the monkey, Mark Twain suggested, only to find that humans were no great improvement.[5]

All these creatures, whether them or us, trod delicately on the Earth, leaving only occasional bones, footprints preserved in ash, and the worked stones that they loved. In contrast, untold creatures dominated the landscape.[6] Some 250 million beaver dams held back lakes across North America before European settlement, compared with less than eight million today. Bison herds widened channels at river crossings and, on the tallgrass prairie alone, luxuriated in 100 million muddy wallows. Hippo trails through riverbanks promoted flood breakouts (dinosaur trackways may have done the same), and burrowing prairie dogs, some 800 million in Texas alone early in the twentieth century, displaced huge amounts of plains soil. Salmon and trout create nests in river gravel, and crayfish shoulder pebbles aside. Charles Darwin celebrated the influence of earthworms on landscape.

Along with toolmaking, the use of fire has been an important skill for more than a million and a half years in Africa and nearly half a million years in Europe.[7] Clever Neanderthals made pitch from birchbark, used fire in hafting implements, and created fireplaces for warmth and hot food, fostering social interaction and extending the length of the day. Across Australia, the first Aboriginal inhabitants may have used fire to manage landscapes, contributing to vegetation and faunal change on the river plains.[8]

Anthropogenic activities would increasingly influence the Earth and its rivers through agriculture, irrigation, and deliberate engineering (Fig. 14.2). Had they known how profound their effect would be, hominins might have had second thoughts about leaving Africa. Or perhaps, surrounded by fearsome cats and crocodiles, they would have shouted, 'Bring it on!'

The sun is going down in a ball of flame over Palaeolake Olduvai. The creature watches the play of fire at Ngorongoro Crater, brighter than the reddest stone tool. Voices sound from the trees. Standing erect, it answers in the subtle musical language of its kind.

Farming at Jericho

The mound of Tell al-Sultan at Jericho in the Dead Sea rift (Fig. 14.3) is surrounded by palm trees and fields. Early inhabitants exploited a spring on the edge of the rift valley and built settlements on the rubble of houses repeatedly destroyed by armies or abandoned to jackals and vegetation. The spring water flows to the River Jordan where, in biblical times, lions came up from the dense riverbank vegetation, the 'jungle of the Jordan', to attack the sheepfolds.

Water is at the heart of everything in the Near East. Water bursts from case-hardened rock walls and dewfall makes crops possible. In the deep gorges of the Sinai Desert, springs trickle down mossy rock faces where cypress trees mark the welcome shade of an oasis. The bedouin live here with their tents and camels, nomadic masters of an inhospitable land. They know the land and its water.

Among the first cities on Earth, Jericho is a crossroads in time at the beginnings of agriculture—the most important way in which humans modify rivers, then and now. The city has been almost continuously settled for nearly 11,000 years, and at times several thousand people occupied a walled enclosure of more than ten acres.[9] With about twice the present rainfall,[10] the early inhabitants grew barley and wheat, grapes and figs, onions and pomegranates, and the city became famous for date palms, first cultivated 7000 years ago in Arabia.[11]

The 1950s excavations that made Jericho famous were directed by pioneering archaeologist Kathleen Kenyon. The walls exposed in the trenches were built from sundried bricks that bore impressions of the

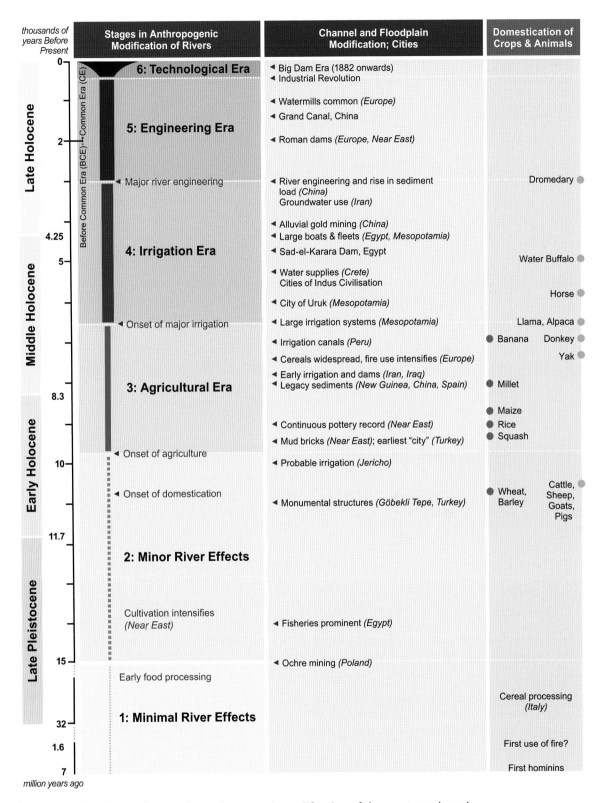

Figure 14.2 Timeline and stages for anthropogenic modification of river systems, based largely on information from the Near East, North Africa, North-western Indian Subcontinent, China, and parts of Europe. The timing and intensity of anthropogenic influence vary greatly regionally, with some stages bypassed completely. After Gibling, 2018.

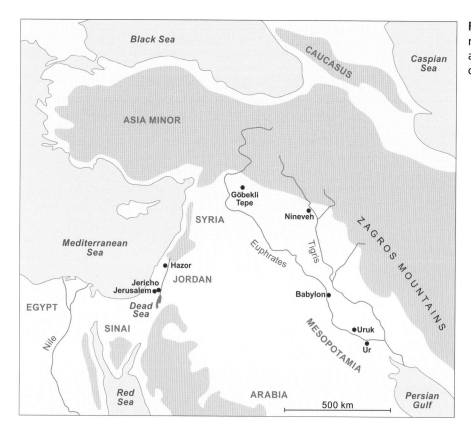

Figure 14.3 Modern rivers in the Near East and archaeological sites discussed in the text.

Figure 14.4 Stone tower in excavations at Jericho on the west side of the Jordan Rift Valley (1979 photo). The tower dates back about 10,000 years, and is one of the oldest known buildings.

brick-maker's fingers, recording an industry that dates back more than nine thousand years at Jericho, and pitted the river floodplains wherever cities grew. Red clay floors were burnished to a polish and in places the weave of mats was visible, one with the track of an ant that had eaten the fabric. And there were the extraordinary plastered skulls that projected from the trench walls or lay under the house floors, the eye sockets containing cowrie shells and hair and moustaches recreated with clay moldings. Were they the remains of enemies or were they, as Kenyon believed, venerated ancestors dispensing wisdom from beyond the grave?

Exposed in a trench on the tell is a tower of roughly hewn stone (Fig. 14.4), a watchtower for flash floods or a link to the heavens. I am looking at one of the oldest city structures, dating back about 10,000 years.[12]

Water runs freely down the Jericho gutters from Tell al-Sultan. I reflect on the random events of the historical record, the finger impressions of a brick-maker, the ant that annoyed a Neolithic housekeeper. It is 1979, and I am on a bus down the Dead Sea road, its loudspeaker blaring (just as randomly) Boney M.'s song *Rasputin*.

Figure 14.5 Schematic diagram to show the range of anthropogenic influences on river systems, especially from about 10,000 to 4000 calendar years BP (Before Present). Figure drawn by Meredith Sadler. After Gibling, 2018.

The rise of agriculture and irrigation

Inhabitants of the Fertile Crescent laid the groundwork for the so-called *Neolithic Revolution* that established agriculture and changed rivers and landscapes for ever (Fig. 14.5). Humans had long collected and processed cereals and other foods, and fish was a staple food for river communities around the world (Fig. 14.6). But the revolution ushered in irrigated farming in many parts of the world between about 10,000 and 5000 years

Figure 14.6 Fishing on the Ganga River near Bithur, India. Archaeological evidence indicates active fisheries at sites around the world from the latest Pleistocene onwards.

Figure 14.7 Terraces in the Himalayan Foothills, Nepal, with harvesting of barley. Terrace systems were generated in the Middle East at least 6000 years ago.

ago, accompanied by the rise of urban civilization. But why farm? Why exchange foraging with its short working week and rich diet for agriculture with its toil, poor nutrition, and crowded living? Climate change, atmospheric carbon dioxide levels, spread of new plant species, overkill of large animals, population pressure, the challenges of nomadic life, and the desire for services provided by craftsmen, administrators, and priests—all have been debated.[13]

Some 14,000 years ago the Natufian culture of the Fertile Crescent made a remarkable leap forward.[14] They gathered in permanent communities that have yielded many stone sickle blades, designed to be hafted onto bone or wooden handles and probably used for harvesting cereals. Were they the first farmers, or did they simply gather wild cereals and clear unwanted vegetation? Although there is little to suggest that they domesticated crops, their emerging culture provided the tools for agriculture.

Domestication of cereal crops that included barley and wheat was underway in the Near East by about 10,700 years ago.[15] Genetic changes in cereals at archaeological sites testify to deliberate selection for higher yields and the retention of grains on the stalk to aid harvesting. Within a short time, cattle, sheep, goats, and pigs—grazers and land-degraders *par excellence*—were domesticated across the region. The onset of agriculture in the region has been dated to 9800 years ago, defined as an organized system that relied heavily

on the production and consumption of domesticated plants. Within a few thousand years, most Near Eastern people became farmers as cereal production spread across Europe and Asia. Crops were widely grown on agricultural terraces (Fig. 14.7), which provide an even surface for soil retention, irrigation, and ploughing on hillslopes. Terraces are thought to have originated in Asia, and some in Yemen date back at least 6000 years.

Some 11,000 years ago, hunter-gatherers at the dawn of the Neolithic assembled at Göbekli Tepe in Turkey.[16] There they danced and feasted, probably drinking beer made from fermented wild crops. Around them stood stone pillars carved with representations of snakes, scorpions, birds, and human hands. Long before writing or even agriculture, they lived in a symbolic and abstract world with a monumental mythology.

Sustained agriculture in seasonal climates requires irrigation. The emergence of irrigation, the use of rivers, and the settlement of large populations on river plains is *the* history of much of humankind, and humans became a powerful geomorphic force in the 'anthropocene'.[17] Pollen records from the bogs of western Europe show that oak and elm yielded to scrubland as grazing and arable land expanded. In parts of Europe, China, and New Guinea, such landscape changes date back more than 7000 years. And *legacy sediments* choked the rivers as shallowly rooted crops replaced the deeply rooted forests, releasing a flood of soil and weathered rock.

The use of underground water

Not all water for agriculture and cities came from rivers. As at Jericho, groundwater and springs played a key role, bringing large volumes of underground water onto the river plains. In addition to shallow wells, an important groundwater system of the ancient Near East was the *qanat*, which originated in Iran some 3000 years ago and spread to the Nile Valley and along the Silk Roads to Asia. Vertical shafts met a gently sloping tunnel as much as 70 km long, which intersected the water table at a high elevation and brought water to the fields below (Fig. 14.8).[18]

By the Iron Age, engineers at the hilltop city of Hazor were exploiting water that gushed up the Dead Sea Fault, excavating a chamber where the inhabitants could descend stairways to the water table.[19] Jerusalem, too, with its porous limestone depended on springs and small creeks. Facing attack from an Assyrian army in 700 BCE, King Hezekiah hastily blocked the water courses that drained from the city, denying water to a besieging army. Showing a clear understanding of underground water, his forces redirected an important spring to flow inside the city walls through the rock-cut Siloam Tunnel, half a kilometre long. The crews worked from both ends, axing through solid rock in a race against time.[20] Soon the tunnel was below the depth where sound could penetrate to guide the tunnellers, and both crews curved their tunnels towards shallower rock to regain acoustic contact. Presently, voices could be heard, and the tunnels broke through with a difference in elevation of only a few tens of centimetres.

From the early centuries BCE to the Roman and Byzantine empires, a society of water exploiters lived across the Near East. Builders of the rock-cut city of Petra, the Nabateans were skilled hydrologists who inhabited deserts without perennial rivers, where the water table was generally too deep for shallow wells.[21] They directed the rare torrential rains that poured down hill slopes into underground cisterns, and they fed runoff into dry riverbeds for cultivation, supporting tens of thousands of people.

At the Western Wall in Jerusalem, caper bushes, adept at exploiting water, thrust out between the huge limestone blocks of Herod's temple. Scraps of paper with prayers project from the crevices. There I too place my requests for this land. My journey now takes a farcical turn. I miss my flight in Athens and, stranded penniless, subsist on bread and water for some days in the forest, before flying on to communist Romania. At the airport, I strike up conversation with a militiaman waiting to meet his father, an oil executive. Drunk and dishevelled, the oilman presents his son with a vinyl record of *Rasputin*. As we drive into town in the militia bus, the son gleefully repeats the final words … 'those Russians'.

Irrigation in Mesopotamia

The river plains of Mesopotamia, the legendary Garden of Eden, are traversed by the Tigris and Euphrates rivers, fed by rain from unstable westerly winds. The crops require irrigation to survive the intense summer heat that follows the spring floods. Mesopotamian history is all about how societies managed or mismanaged their rivers and land as they developed from rainfed civilizations to riverine civilizations and then to hydraulic civilizations.[22]

More than seven thousand years ago, barley and wheat flourished across the plains, supporting tens of thousands of city dwellers.[23] From this period dates the first recorded picture of a plough,[24] used by farmers who also built large-scale irrigation and flood-control

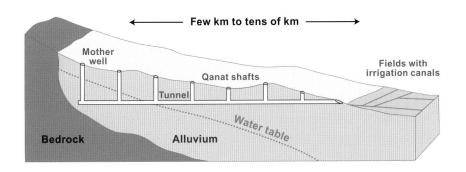

Figure 14.8 Schematic diagram to show construction of a Qanat system. The tunnel brought water from areas with a high water table to areas with a low water table downslope, allowing irrigation of fields. After Lightfoot, 1996.

systems. Even earlier in the nearby uplands, farmers may have constructed the first large dams, diverting water into irrigation canals behind ramparts of brushwood and earth.[25]

By 6000 years ago, the plains were criss-crossed with irrigation canals.[26] The famous code of Hammurabi, dating to about 1750 BCE, frowns upon farmers who neglect their dikes and flood a neighbour's land, although Hammurabi himself diverted channels to destabilize a competitor city downstream. While encouraging these powerful civilizations, the fickle rivers (in the words of scientist Galina Morozova) created and destroyed towns, fomented wars, and fuelled the pessimism of unstable populations.[27]

The Assyrian rulers were justly proud of their river engineering. 'I constrained the mighty river to flow according to my will,' declares an inscription on the tomb of the legendary Queen Semiramis, 'and led its waters to fertilize lands that before had been barren and without inhabitants'.[28] King Ashurnasirpal boasted:

I dug a canal from the Upper Zab River; I cut straight through the mountains … I provided lowlands along the Tigris with irrigation … pomegranates glow in the pleasure garden like the stars in the sky … in the garden of happiness they flourish like cedar trees.[29]

Irrigation also had damaging effects. It raised the water table, and salt crystals precipitated in the soils in the summer heat, reducing fertility by about 2400 BCE and requiring farmers to replace salt-sensitive wheat with barley. The destruction of forests that included the famous cedars of Lebanon contributed legacy sediments to the rivers.[30] And irrigation-based societies were not immune to drought, which at times caused civilizations to decline more than 8000 years ago.[31]

With his sharp eye for rivers, Herodotus described how the rulers of Mesopotamia in the first millennium BCE used a vast workforce to engineer the river lands.[32] When one of his sacred white horses drowned in a Tigris tributary, Cyrus the Great of Persia swore that he would punish the river and make it so weak that a woman could cross without wetting her knees. His hapless soldiers wasted their summer dividing the unfortunate stream into 360 feeble channels. Resuming the attack on Babylon, Cyrus diverted the Euphrates and lowered the

water to thigh depth, enabling his troops to storm the city's defences along the channel's entry and exit points.

The *Epic of Gilgamesh*, dating back at least 4000 years in written form, drew on the rivers of Mesopotamia to illustrate human vulnerability and impermanence. Gilgamesh king of Uruk bemoans his short and mean existence: 'Here in the city man dies oppressed at heart, man perishes with despair in his heart. I have looked over the wall and I see the bodies floating on the river, and that will be my lot also'.[33]

Living with the Nile

Few countries are as dependent on a single river as Egypt is on the Nile. Although the Nile is a predictable river with a nearly optimal yield of water and soil nutrients, Pharaohs and labourers alike would have perished had the flood failed even for a season. Watchers recorded the river level using *nilometers*, gauges inscribed in stone walls. But, because the river's source was unknown, its flow could only be monitored, not controlled, and was a matter of prayer or fatalism.[34]

As the African monsoon intensified about 15,000 years ago following the Ice Age, the Sahara Desert supported large lakes and a substantial human population. But a decline in precipitation about 5000 years ago rendered large areas of North Africa uninhabitable.[35] Now everything depended on the Nile.

Set between upland margins, the Nile floods much of its valley and the floodwaters drain back into the river.[36] Herodotus recorded the practice of recession agriculture as the floods receded, when the Egyptians sowed their plots and turned pigs loose to tread in the seeds. It was an advantage that the autumn flood inundated the floodplain after the summer heat had burned off the weeds. Unlike Mesopotamia, long irrigation canals were impracticable, and the Egyptians guided the floodwaters with artificial levees and short canals, using earth dikes to trap floodwaters and nutrient-rich silt. A ceremonial macehead dating back more than 5000 years shows the Scorpion King cutting an irrigation ditch by a waterway that branches towards irrigated fields.

About 4600 years ago when the first pyramids were being constructed, Egyptian engineers built the world's oldest large dam, the *Dam of the Pagans*, to provide water for quarrying.[37] The earthen dam was 14 m high and more than 100 m long, faced with 17,000 cut stone blocks. But not all river works had such a high purpose. Herodotus mentions Queen Nitocris of Egypt who, seeking revenge for the murder of her brother, constructed a large underground chamber and invited to a banquet everyone implicated in his death. When the party was in full swing, she let in the river through a concealed pipe.[38]

As irrigation expert Daniel Hillel notes, the Nile's dependable flood regime and the rich silt transported by the Blue Nile from Ethiopia 'enabled Egyptian farmers to produce a surplus that fed the artisans, scribes, priests, merchants, noblemen, and—above them all—the Pharaohs who used their coercive power to order the building of self-aggrandizing monuments. Those monuments still stand today, less in testimony to the vainglorious kings who ordered them than to the diligence and organization of a society of labour nurtured by the river and rooted in the land irrigated by it.'[39]

Early agriculture in the Americas

In the Americas, agriculture also developed early.[40] In Mexico and the Andes, maize and squash were cultivated and processed nearly 9000 years ago, and in the northern Andes, squash was domesticated some 10,000 years ago and peanuts nearly 8000 years ago. Peruvian dryland farmers developed sophisticated terrace systems and water-control structures, and deforestation linked in part to maize cultivation triggered landscape erosion in Central America more than 3000 years ago.

Amazonia in South America has been considered a near-pristine wilderness with minimal human modification. However, early adventurers described large settlements along the river bluffs, and at least eight million people may have been living in the region by 1492. Widespread anthropogenic soils in the river settlements (see Chapter 20) suggest substantial early 'landscape domestication', and Amazonia boasts at least 83 domesticates that include the root crops manioc and sweet potato, peanuts, cacao, tobacco, pineapple, chilli peppers, brazil nuts, and fruit trees. Maize and squash were in widespread use some 6000 years ago, and farming was widespread by 4000 years ago, supported by canals, cuts between river meanders, and fish weirs. At one site, a shell mound or midden some six metres high, dating back more than 9000 years, shows that wild rice (*Oryza* sp.) was domesticated about 4000 years ago.

In North America, early farming that included maize cultivation was also widespread. On the Gila River, a Colorado tributary in Arizona, indigenous communities built an earth dam that was five kilometres long, with hundreds of kilometres of irrigation canals under a carefully engineered gradient. Large areas were irrigated nearly 3000 years ago.[41]

Rice and irrigated ricefields

More than 9000 years ago in southern China, a group of hunter-gatherers took one of the most important steps in human history.[42] Selecting wild rice, *Oryza rufipogon*, a swamp plant that had long supplied them with grain, they began a deliberate process of cultivation that involved the modification of hillslopes and floodplains to hold standing water. Genome analysis of wild and cultivated rice suggests that the innovative villagers lived along the Pearl River or Zhujiang inland from Hong Kong. They sought out plants with stronger and straighter stems, heavier and more abundant grains, and grains that clung to the plant rather than falling to the ground.

Rice was a brilliant success. By 8000 years ago rice cultivation had moved north to the Yellow River plains where Neolithic farmers were already exploiting millet,[43] and river clays at least 4000 years old in the Loess Plateau contain evidence for rice farming, with water-laid pottery and charcoal introduced during field manuring. By 6000 years ago, the use of domesticated rice was widespread across China, as indicated by archaeological sites with paddy fields and irrigation systems choked with rice husks, stems, and leaves along the Yangtze River.

Rice farming spread through Southeast Asia (Fig. 14.9A) and reached the Indian subcontinent at least 4000 years ago. In Japan, rice was considered the most sacred thing on Earth after the Emperor. Moving west along the Silk Roads, rice reached Mesopotamia and the Nile Delta in the early centuries BCE,[44] transported by horses, donkeys, and yaks (domesticated more than 6000 years ago) and camels (domesticated some 3000 years ago).[45] Wherever it was farmed across Eurasia, rice transformed river landscapes and navigation (Fig. 14.9B). Apparently independently, wild rice was domesticated in Amazonia 4000 years ago, and wild rice (*Zizania*) was of great importance for some Native American communities.

Figure 14.9A Rice fields on the floodplains of the Chao Phraya river system, Thailand. Note the long pole of the well sweep (shaduf), for lifting water for irrigation or other uses.

Figure 14.9B River boats on the Chao Phraya River system, Phitsanulok, Thailand, for transporting rice to the delta. Riverbank infrastructure associated with water transport and improvements for navigation would have contributed to river modification during the earlier Holocene.

Figure 14.10 Creek at Chopani-Mando, a tributary of the Belan River of India, cut into Proterozoic quartzites, an extension of the Kaimur Hills in the distance. The far bank was the site of a Paleolithic to Mesolithic settlement, with artefacts scattered on the terrace. A nearby Neolithic site yielded evidence for domestication of rice, among the earliest such sites in India.

Deserving the eternal gratitude of Asian rice farmers is the water buffalo, which was domesticated more than 6000 years ago.[46] With its large feet and love of water, the gentle buffalo ploughs comfortably through the mud. Small boys dive off the animals as they wallow in the rivers and ride the buffalo home with a blaring transistor radio hung on a horn.

Early rice cultivation in India

The road to the Belan River begins in Allahabad on India's Ganga Plains. In 1980, Professor Sharma and his team from Allahabad University documented some of the subcontinent's earliest rice cultivation in *Beginnings of Agriculture*,[47] unveiled at a ceremony attended by Prime Minister Indira Ghandi.

On this early January morning, a thin mist swirls across the confluence of the Yamuna and Ganga rivers, where a huge tent city is loud with traditional music. Every twelfth year, more than 60 million people gather here for the festival of Kumbha Mela, the largest human assembly on Earth. Nothing is a pristine landscape: the agricultural Ganga Plains are among the world's most densely peopled regions, a testimony to the enduring power of the Neolithic Revolution. But the villages are largely self-sufficient, leaving little more of a footprint than their Neolithic forerunners. Soon the rugged Kaimur Hills rise into view to the south, and the road descends to the Belan, wide and shallow in the dry season.

The boatman squats on the stern of the boat, paddling timelessly forward. Birds fly up without haste—a black-winged stilt with its red legs, a hoopoe with its curved bill and crest. The people along the banks are unhurried and dignified. The boat drifts to a stop by a rocky channel where the riverbank is littered with tiny translucent blades and a faceted corestone (Fig. 14.10)—precision tools made by capable Paleolithic and Mesolithic peoples using quartzite, carnelian, and jasper from the Kaimur Hills.

On the bank stands a stately elderly man, wiry and grey-haired. He is Ramkhemawa who oversaw Professor Sharma's excavations. Patiently he points out the foundations of huts littered with scrapers, anvils, and hammer stones, as well as primitive pottery that predated agriculture. Rice grains from the excavations suggest that the inhabitants, probably Kaimur hunter-gatherers, collected wild rice or, perhaps, experimented with cultivation. Downstream, the boat comes ashore by a ravine strewn with younger Neolithic debris. Here the archaeologists unearthed huts and a clay floor imprinted with hooves, as well as pottery sherds tempered with rice husks. The rice was almost certainly cultivated and processed in the village, implying domestication.[48]

The boatman brings back a honeycomb from a gully, an association between farmers and the honeybee that dates back more than 8000 years.[49] Along the dusk road to Allahabad, a stream of humanity appears in the car headlights—women carrying firewood and children on heavily loaded bicycles too large for them. It is rice and curry for dinner at the university guesthouse. Rice, the modifier of river plains, and food for half the world's population.

Roman water engineering

During the last centuries BCE and the start of the succeeding millennium, the Romans took over much of Europe, North Africa, and the Near East. They developed a formidable engineering civilization with rivers and water as a central part of their strategy. My introduction to the Roman Empire came through high-school Latin and Julius Caesar's *Bello Gallico* (*Gallic Wars*), which contained constructions such as 'Caesar, winter quarters having been set up'. What cool competence! What sublime power! His attempted conquest of the British Celts would have been considered a failure had we any account but Caesar's own. Caesar is never wrong, and when he praises an enemy you know that victory will follow.

Roman cities with their water supplies from rivers and springs became models of habitation (Fig. 14.11).[50] In his book *Agricola*, Cornelius Tacitus, who believed in Roman destiny, cynically mocked the hapless Britons for adopting Roman civilization—poor, deluded slaves whose valour perished with their freedom. Public water was the measure of civilized living, and the high water mark of a city's magnificence was its luxurious baths, heated from wood-fired furnaces where, at the end of

Figure 14.11 Roman city of Pompeii, which boasted an elaborate water system. The ruins include the remains of houses excavated from volcanic ash and cart ruts on volcanic cobbles in the street. The city was destroyed in 79 CE by the eruption of Vesuvius, the flank of which is visible in the distance.

the business day, relaxed Romans could take massage and pursue business contacts.

Ancient Rome used aqueducts nearly 100 km long to supply a population that, by the end of the first century CE, may have exceeded half a million.[51] The emperor personally took nearly 20 per cent of the water, and waste was flushed into a major sewage system, the *Cloaca Maxima* or 'Great Drain' with an outlet to the River Tiber, perhaps constructed as early as the seventh century BCE and considered a wonder of Rome (Fig. 14.12). By the fourth century CE, Rome boasted several thousand public baths, fountains, and basins, along with two artificial lakes (*naumachia*) where, in a ghastly form of mass entertainment begun by Julius Caesar, thousands of condemned prisoners fought to the death in naval battles. Deforestation and irrigation intensified, and the Po delta advanced far out into the Adriatic Sea, charged with river sediment washed from the fields and uplands.[52] Especially in Spain, France, and the Near East, Roman military engineers developed large hydraulic systems that supplied food and water for the occupying legions and intensified agriculture in the empire's last days. Some forty-five large dams were built across rivers in the Near East, and aqueducts and *qanat* systems supplied cities and farms.[53]

During the medieval period in Europe, astute engineers set the foundation for modern river and water supplies, often drawing on a knowledge of Roman systems. Under the cobbled streets of Siena in Italy run many kilometres of medieval, brick-walled tunnels (*bottini*) that channel the rainwater and surface flow

Figure 14.12 Exit of the Cloaca Maxima, the main effluent drainage system of Ancient Rome to the Tiber River. The early drainage system dates back to the 7th century BCE. 'The Mouth of the Cloaca Maxima (I)', by George W. Houston.

percolating down through the porous rock. Branches supply residents, some still named on the walls, and on the still surface of the water float rafts of precipitated calcite, suspended by surface tension. After a long subterranean journey, grimy geologists led by city resident Vincenzo Pascucci rush out to the fountain in the square, emerging through a posh restaurant where the well-heeled diners are considerably startled.

CHAPTER 15

THE LOST SARASWATI RIVER OF THE INDIAN SUBCONTINENT

Saraswati, the Mother of All Rivers

Thousands of years ago, so it is said, a group of travellers left the plains of Central Asia.[1] Traversing the Khyber Pass or the Bolan Pass near the Helmand River of Afghanistan, they descended to a marvelous land, prosperous beyond the dreams of a desert people—the Land of the Seven Rivers, traversed by the Indus and its tributaries from the uplands and glaciers of the Himalaya. The river plains were occupied by peoples with a long history, but the travellers kept alive their cultural traditions, including the use of horses, the

lightweight spoked wheel, and the chariot. For plains people used to lumbering oxcarts with solid wheels, the streamlined form and speed of the chariot was an exciting novelty. And with the spoked wheel came a curious concept—the wheel of time with its lines and divisions.

The travellers, it is thought, developed an oral tradition more durable than crumbling brickwork. Many years passed before the tradition was committed to writing as the *Rigveda*, an anthology of poems of uncertain

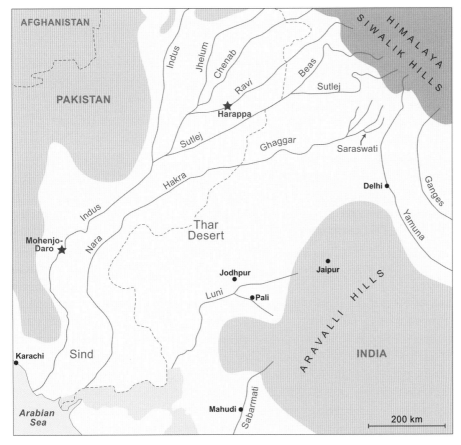

Figure 15.1 Water courses associated with the legendary Saraswati River of the north-western Indian Subcontinent.

date that celebrate earthly existence. The Vedic authors knew desert lands and river crossings, as well as the elephants and peacocks of the plains. But cities are not mentioned nor even brick, although there is reference to mud-walled forts like those of the Himalayan passes.

This is one version of a disputed history, for the Vedic writers recorded their observations as imaginative accounts. Did they truly come from 'away', or were they indigenous people who used new concepts to create traditions? Were new ideas brought by travellers or by victorious armies? And what, if any, is the link between the Vedic tradition and the early cities of the Indus Civilization?

What is certain is that the Vedic peoples loved and respected rivers. I pick up the *Rigveda* and, divorced in almost every way from the unknown writers, I feel the wonder of the rivers as they knew them thousands of years ago. And no river is more exuberantly described than the Saraswati—a river deified as the goddess of music and the arts, the consort of Lord Brahma the creator.

But where *was* the Saraswati? Today, the name belongs to an inconspicuous tributary of the Ghaggar River of northern India (Fig. 15.1), itself a shallow ephemeral river that flows primarily during the monsoon. Could such a minor channel have inspired the Vedic writers? Nevertheless, the river's general location is recorded in the *Rigveda* in order among the rivers of the Punjab—the Indus, Jhelum, Chenab, Ravi, Beas, Sutlej, and Saraswati, west of the Yamuna and Ganga. Was the Saraswati the great Indus itself, a 'river of rivers', capable of loosening rocks from the mountain sides as readily as one digs up lotus roots? Was the name a transliteration of the Helmand River? Or was there once a powerful Saraswati that flowed to the sea but was later reduced to a backwater or lost in the desert sands?

Max Müller translates the Rigveda

Friedrich Max Müller had a rough start in life. His father, the German soldier and poet Wilhelm Müller, died when Max was four, and the family struggled on the edge of poverty.[2] Max's breath froze at night on the bedsheets. He attended school in Leipzig and, a gifted pianist, frequented the house of composer Felix Mendelssohn. By the age of twenty he had received his

doctorate in Sanskrit, the ancient language of the Indian subcontinent, which had become known to Europeans in the seventeenth century.

Müller moved to London to examine Sanskrit manuscripts of the *Rigveda* owned by the East India Company. Narrowly surviving the 1848 revolution in Paris, he lived in Oxford for the rest of his life. His translation of the *Rigveda* was a masterpiece, and he considered the writings to represent the earliest surviving expression of spirituality, at the dawn of religious consciousness. Rumours spread that Müller had discovered a strange, ancient religion, and he gave lectures to enthusiastic crowds that included an attentive Queen Victoria. His death in 1900 brought condolences from the Queen, and an Indian colleague wrote to Müller's wife that all of India was mourning with her.

The *Rigveda* writers loved their rivers, and verses in praise of the Saraswati are considered among the earliest parts. The roar of the flood water resembles a bellowing bull, and the torrential flows are likened to horses, chariots and, for their nurturing properties, cows. And the coming of the monsoon is beautifully portrayed:

> The music of the frogs comes forth in concert like the cows lowing with their calves beside them. When at the coming of the rains the water has poured upon them as they yearned and thirsted, One seeks another as he talks and greets him with cries of pleasure as a son his father … So, frogs, ye gather round the pool to honour this day of all the year, the first of Rain-time.[3]

River courses and the Rigveda

The British loved India. Not so much, perhaps, the unfamiliar cultures, but the land and empire that they gradually acquired. Much of the Indus basin came under British control during the 1840s, when expeditions sought a road through the Thar Desert, widely known as the 'region of death'. To their surprise, military commanders discovered dried-up water courses and villages with wells and oases, an observation that had not escaped earlier armies marching in from the west.[4] Colonial engineers began a systematic construction of canals that extended earlier irrigation systems, and soon the newly available river maps began to suggest intriguing ideas about the region's history.

Armed with Max Müller's translation of the *Rigveda*, scholars attempted to link the mythical Saraswati to a real water course[5]. C.F. Oldham was Surgeon Major in the East India Company, and published accounts in 1874 and 1893. His geologist cousin R.D. (Richard Dixon) Oldham published his research in 1886, later speculating on Himalayan river histories (see Chapter 6). C.F. noted that the *Rigveda* describes the Saraswati as a majestic river flowing from the mountains to the sea, implying a High Himalayan connection, whereas later texts describe the river as losing itself in the sands. Rising in the Siwalik Hills, the modern Ghaggar and Saraswati are not connected to the High Himalaya with its glacial meltwater, and C.F. suggested that they had always been small rivers fed by the monsoon rains. R.D. pointed out that the Ghaggar runs into a largely abandoned river named the Hakra upstream and the Nara downstream, which traverses the Thar Desert to the Arabian Sea. Was this the Saraswati? Was it the former course of the Yamuna, which later shifted east to join the Ganga, or was it the former course of the Sutlej, which later shifted west to join the Indus?

Both men linked the Saraswati to the Sutlej, which shifted west to leave a small modern 'Saraswati' in a channel far larger than its needs. Neither Oldham gave much credence to dwindling rainfall to explain the Saraswati's decline. Had this been the case, thought R.D., all rivers in the region should have diminished. Heavier rains in earlier times, declared C.F., would have required 'such meteorological conditions as must have rendered the holy land of the Brahmans an uninhabitable swamp'.[6]

By the late nineteenth century, the *Rigveda* account had been vindicated. The Saraswati had existed and its impressive floods suggested that its headwaters lay in the High Himalaya, implying a link to the Sutlej, the Yamuna, the Indus, or variously all three.[7]

The Indus Civilization and the Saraswati

In 1924, John Marshall, Director-General of the Archaeological Survey of India, announced that a previously unknown urban civilization had been unearthed from tells at Harappa and Mohenjo-Daro in the Indus region. The excavations had been carried out by archaeologists Rai Bahadur Daya Ram Sahni and R.D. Banerji a few years earlier. Writing in *The Illustrated London News*, Marshall noted that tells were concentrated along rivers where the city dwellers would have found a reliable water supply.[8] The news article included photos of the mounds with their brickwork and, at Marshall's request, photos of seals bearing mysterious pictographic symbols in case readers could decipher them (the symbols remain undeciphered).

The archaeologists had discovered the Indus Civilization, among the most important Bronze-age urban cultures. At a stroke they had restored several thousand lost years to the subcontinent's history and rekindled the Saraswati debate. The cities had monumental structures and Mohenjo-Daro may have been the largest city of the ancient world, supported by plough-based agriculture, hunting, and plant gathering. The culture was thriving more than 5000 years ago, at about the same time as cities in Egypt and Mesopotamia, but the major settlements were largely deserted within a short time.

The citizens of the Indus Civilization revelled in the luxury of available water. There were many wells, enormous public baths, and houses with terracotta conduits for wastewater, as well as street drains and sewage pits. No other civilization showed such skill until Roman times. Most remarkable of all, no glorification of warfare was evident, and despite an impressive civic order, its rulers were unknown.

Earlier travellers were familiar with the tells. British engineers had used bricks from Harappa for railroad tracks,[9] and C.F. Oldham and the later adventurer Marc Aurel Stein noted the many brick mounds along the Hakra, which local inhabitants attributed to a once-great river. More than two thousand sites occupy an area of about a million square kilometres, greater than that of ancient Egypt and Mesopotamia combined. But the greatest density lies in the drainage divide between the Ganga and Indus systems, far from large modern rivers.

Were towns of the Indus Civilization founded on the banks of the vanished Saraswati? Did their citizens write the *Rigveda*? And what ended the civilization?[10] Some have invoked overgrazing and a demand for fired bricks that depleted the riparian vegetation. Others thought that the rivers broke through the natural levees or were diverted by earthquakes to leave the cities waterless. But excavated seals depict tiger, buffalo, rhino, and elephant—animals of the humid forests—with few images of the desert camel. Had dwindling monsoon rains caused the Saraswati to dry up?

A waning monsoon and declining rivers

Far to the south of Harappa, the village of Mahudi stands on a cliff where a road angles down to the Sabarmati River, crossed by camel-carts laden with brushwood (Fig. 15.2A). Dry-season crops cover the sand bars, and curious villagers and camels peer over the shoulders of the geologists.

The topmost layers of sand in the cliff were laid down 12,000 years ago, at a time of vigorous winds in the Thar Desert, after which powerful river flow from monsoon rains must have cut through the strata to form the cliff (Fig. 15.2B). The sand bars in the modern river are less than 5000 years old, dating from a time when the river could no longer transport its sediment. Thus, a vigorous Sabarmati River cut the cliff less than 12,000 years ago, but by 5000 years ago the monsoon was declining.[11]

A similar story emerges from the nearby Luni River, a wide dry-season expanse of sand where vultures surround the carcass of a water buffalo and buffalo horns project from an animal sunk upright in the sand. In the early nineteenth century, British officer James Tod noted the dangerous quicksands of the 'salt river' where the camel caravans crossed.[12] As at Mahudi, the river cliffs were cut less than 12,000 years ago, and a nearby lake was fresh about 10,000 years ago, supplied with nearly twice the present rainfall, but dwindled to a saline pan 5000 years ago.[13] Thus, over a few thousand years, the region's rivers dwindled from powerful torrents to seasonal flows as the Thar Desert expanded. Perhaps a waning monsoon had reduced the Saraswati to a trickle.

Satellites find the Saraswati

Satellite images finally began to resolve the Saraswati question by linking the channel of the Hakra, more than 10 km wide, to the Sutlej downstream from the big river's mountain exit.[14] Researchers manoeuvred drilling rigs out to the channel line where they penetrated a thick layer of river sand charged with fresh

Figure 15.2A Sand and gravel in cliffs near Mahudi, with luminescence dates of about 300,000 years for the lowermost strata and 12,000 years for the topmost strata (out of sight).

Figure 15.2B Cliffs 40 m high along the Sabarmati River at Mahudi, northwest India. Luminescence dates of 12,000 years ago for the topmost cliff strata and 4500 years ago for the sandbars suggest that the monsoon intensified during that period, when increased river discharge caused incision. Subsequent waning of the monsoon promoted sediment buildup in the channel.

Figure 15.3 Festival gathering on river bars at the confluence between the Ganga and the Yamuna at Allahabad, India.

groundwater, with grains that matched those from the Sutlej headwaters. The topmost sand and mud had been laid down in narrow channels, dwarfed within the broad valley and attributed to the seasonal Ghaggar. Although a waning monsoon reduced river flow and the loss of some Yamuna drainage may have played a part, R.D. and C.F. Oldham had been correct in linking the Hakra primarily to the Sutlej. The big Himalayan river had once crossed the flat watershed between the Indus and Ganga but had switched sharply to the west, perhaps tilted by an earthquake or diverted by large floods.

At first sight it seemed that the Sutlej was the mother of all rivers praised in the *Rigveda*. However, age dates showed that the Sutlej abandoned its Hakra course more than 12,000 years ago, long before the towns of the Indus Civilization were founded. Seemingly it was not the river's *presence* that encouraged settlement but its *departure*.[15]

And, after all, is this so surprising? A Himalayan river in flood is a wonderful spectacle but a dangerous ally. A strong foothills river fed by seasonal rain is a lifeline, flowing within a broad valley suitable for farming and settlement. The region had been an agricultural centre in historic times, and traveller Ibn Battutah had praised the region's rice harvest in the fourteenth century. Supplementing modest river flow, a little rain may sustain agriculture on dry-season sand bars like those of the Sabarmati. And the fresh water recharged into the ground by the former river is a farmer's best friend.

The modern Saraswati is a disappointing trickle late in the dry season, hardly the mother of all rivers. No river looks its best under these conditions, but this is not the torrential river that almost drowned Rudyard Kipling's Kim among the dancing boulders.[16] But the glory of the ancestral Saraswati would have passed out of knowledge without the Vedic writers, who were more tightly bound to their landscape than we can possibly imagine. Their patterns of life have also survived. I think of the huge gathering at Allahabad where the Ganga and Yamuna meet (Fig. 15.3) and where, in a later tradition, the Saraswati flows invisibly underground to a triple confluence.[17] The Saraswati 'whose limitless unbroken flood, swift-moving with a rapid rush, comes onward with tempestuous roar … she with her light illuminates, she brightens every pious thought'.[18]

CHAPTER 16

CONFUCIAN ENGINEERS ON THE YELLOW RIVER OF CHINA

Marco Polo explores the Yellow River

In the winter of 1128 CE, the Chinese city of Kaifeng came under threat from Jin cavalry across the empire's northern frontier.[1] As the Jin swept south, Governor Du Chong deliberately breached the dikes along the Yellow River, receiving five promotions in six months before going over to the enemy. The Jin established their capital at Beijing but were swept away by the Mongol hordes of Genghis Khan. By 1279 his grandson Khubilai Khan was ruling a large and populous empire, and trade revived along the Silk Roads. But the Yellow River stubbornly refused to reoccupy its former course. Instead it embarked on an invasion of its own along a route to the Yellow Sea 500 km south of its former course.

The Yellow River came to European attention through the writings of Marco Polo.[2] Born to a merchant family in Venice in 1254, Marco was fifteen when his father Nicolò and uncle Maffio, long presumed dead, returned unexpectedly from the court of Khubilai Khan. Disgusted by Maffio's filthy Mongol coat, his wife handed it to a passing beggar, unaware that a fortune in jewels was sewn into the lining. Pursuing the beggar to the Rialto Bridge, Maffio feigned madness and retrieved the coat. Faced with friends who doubted their exploits, the brothers appeared in their Mongol robes at a party and slit open the hems to release torrents of jewels, after which the neighbours were more attentive.

Two years later they returned to China with Marco along the Silk Roads. Skirting the Taklamakan Desert, they reached Lanzhou, noted for its waterwheel innovations (Fig. 16.1A,B), and followed the Yellow River around the Ordos Loop to Beijing. Marco lived in China for most of the next two decades, part salesman and part dazzled courtier. He was sent by the emperor on a fact-finding mission up the Yellow River to Chang'an (now Xi'an), and he observed silk production and paper

Figure 16.1A Yellow River at Lanzhou, northwest China.

Figure 16.1B Traditional waterwheel and water-supply conduits at Lanzhou, known as the 'Waterwheel Metropolis' since the Ming Dynasty in the 15th century CE.

currency with the keen eye of a Venetian trader. He recorded that the river entered the sea through wide channels with an inland station for 15,000 ships.

Descending from the Tibetan Plateau, the Yellow River traverses the Loess Plateau with its thick

Figure 16.2 Loess of late Cenozoic age in the upper part of a cliff near Lanzhou. The largely structureless, silt-sized yellow loess is up to 400 m thick in the region and rests on a thin gravel-covered river terrace (brown layer at mid-cliff level), above older red strata.

Chinese emperors manage the Yellow River

The earliest Chinese emperors grappled with flood control, irrigation, and silt buildup along the Yellow River in the Loess Plateau and across the North China Plain (Fig. 16.3). Some 4000 years ago, hydraulic engineer Yü the Great deepened the channels and carved passages through the mountains, expelling unruly snakes and dragons.[4] So dedicated was he that he passed close enough to his home to hear his wife and children crying but continued on his way. Enabling the flood waters to find their natural course, the heroic engineer established his fitness to rule.

Large dikes and irrigation projects were constructed along the Yellow River in the first millennium BCE. The Zheng Guo Canal near Chang'an was especially famous. Sent to the warlike Qin to distract them with a debilitating irrigation scheme, water engineer Zheng Guo was unmasked and threatened with execution. Claiming that the canals would establish Qin prosperity for ten thousand generations, he saved his life and, as he predicted, the irrigation scheme enhanced Qin power. Qin emperor Shi Huangdi was buried at Chang'an with his famous terracotta army and, allegedly, a scale model of the empire's rivers and lakes that used a flow of mercury.[5]

With more than 10,000 labourers, Emperor Wu Di constructed the Dragon's Head Canal, named for the discovery of a 'dragon bone', probably a large fossil. The

accumulations of windblown yellow silt (Fig. 16.2). The soft silt pours into the channels from acres of eroding gullies and creeks, and the river transports the sediment to the North China Plain, where through time more than a trillion people have grown crops on the fertile soil.[3] Their self-sacrifice and that of the Chinese hydraulic experts who struggled to contain the river forms a backdrop to the most remarkable river engineering in human history. But in the end, even the most effective emperor could not control such an unpredictable river.

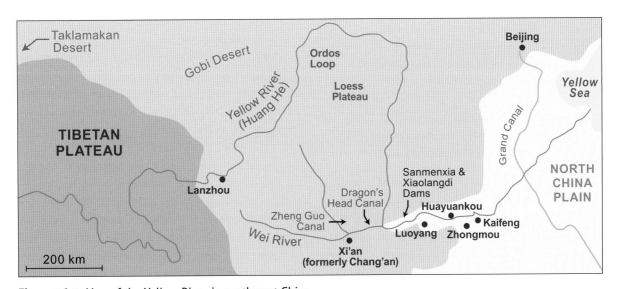

Figure 16.3 Map of the Yellow River in northwest China.

Figure 16.4 Grand Canal in Suzhou, China, nearly 1800 km long. The canal was completed in 609 CE to link the rice-growing south and millet-growing north of China, across climatic zones. Shutterstock©PhotoByLola.

emperor rebuilt a dike that failed in 132 BCE, sending 700,000 refugees adrift, blocking the breach with bundles of branches and earth and composing a psalm after throwing a jade ring into the river and sacrificing a horse:

> The river broke through at Huzi;
> What could we do? …
> The river raged from its boundaries,
> It has left its constant course.
> Dragons and water monsters leap forth,
> Free to wander afar.
> Let it return to the old channel
> And we will truly bless the gods.[6]

Especially remarkable was the Grand Canal (Fig. 16.4), almost 1800 km long. Linking pre-existing waterways, the canal was completed in 609 CE by a conscripted labour force of five million. Barges could now travel from the Yangtze ricefields to the Yellow River millet fields, upstream to Kaifeng, and up the Wei to Chang'an. In a procession that was more than 100 km long, Emperor Yangdi and his cortege travelled in dragon-boats pulled by thousands of servants. Later emperors considered management of the Grand Canal and Yellow River as the key to imperial power, and the grain fleet floated into the river through lock systems that were invented at the end of the tenth century.

But after nearly a thousand years of stability, the Yellow River was beginning to break out early in the second millennium CE. Emperor Shenzong insisted that the river's natural course be respected, but river management soon required large dikes and the deployment of 200,000 tons of protective brushwood bundles.[7] The Zheng Guo Canal was also silting up, and new feeder canals and dams of woven shrubs stuffed with stones were constructed as engineers used local springs to stabilize the water supply. By the early 1700s, the canal was clogged with mud and embankments were collapsing, exhausting both the engineers and 100,000 weary labourers working in a precipitous gorge.[8]

Confucian river engineers

Water has been central to life and thought in China for thousands of years.[9] Many legends allude to the early challenges of managing rivers, and water has shaped language and political and social thinking: how water surmounts obstacles, when it is turbulent or still. An enlightened ruler channels human nature as rivers channel water, and philosophers and poets drew on water in an extended metaphor that far transcended literary constructs.

Great rivers are dragons, providing water but liable to break out in flood. Should the river dragons be respected and channeled into their natural flow, as Yü the Great recommended? Or should they be reshaped for the common good, as Confucian thinkers proposed?

K'ung-fu-tzu or Confucius was born in 551 BCE. An influential teacher, he stressed proper conduct and character, and he encouraged good government through talent rather than status.[10] Hereditary rulers should delegate power to able and virtuous officials, he thought, and the right to govern was confirmed by peace and harmony. Linked to this philosophy was the Mandate of Heaven, which implied that the ruler governed with Heaven's consent—a consent that could be revoked should corrupt or unjust behaviour release natural disasters. With floods among the worst disasters, river regulation had a significance beyond channel control.

A century after Confucius, hydraulic engineer Ximen Bao is credited with halting human sacrifice, during which a young girl was married symbolically to the Lord of the River, eventually drowning.[11] Attending a ceremony with his soldiers, Ximen Bao expressed outrage that the girl was insufficiently beautiful, and instructed the village elders to inform the Lord of a ceremonial delay. The soldiers threw the elders one after another into the river. When none returned from their missions Ximen Bao left, and no one dared resume the practice.

By the time of the Song Dynasty (960–1279 CE), a classless Confucian bureaucracy was largely in the hands of a civil service based on talent. Candidates wrote exams in isolated booths in a large hall, and to prevent cheating, they might be asked to shoot an arrow at a list of questions, answering whichever was transfixed (Confucius had liked archery). By the time of European contact, Confucian bureaucrats were widely considered rule-driven 'mandarins', but capable river engineers came from the ranks of these educated, ethically trained leaders.

But the hydraulic engineers frequently found themselves in a managerial nightmare. A training in literature and ethics was one thing, but controlling the Yellow River was quite another. They learned river management from subordinates whose technical knowledge exceeded theirs. And always on their minds was the erratic river, which could sweep away their lives and careers in a moment along with the embankment and its workforce. Nevertheless, many river engineers were neither timeservers nor lackeys, and those who really knew the river achieved a god-like status.

Especially famous was Ming Dynasty engineer Pan Jixun, who in 1578 designed a management plan for the Yellow River.[12] Lacking formal training, he talked with river managers and peasants, convinced that silt rather than water was the problem:

> When the sediments are scoured away, the river becomes deep, with fathoms of water passing above the bed of the channel … One builds dikes to confine the water and one uses the water to attack the sediments.[13]

Pan Jixun reconstructed the entire Yellow River defence system of the North China Plain, building inner dikes to confine and deepen the channel so that the rapid current could carry away the silt (Fig. 16.5). Outer dikes contained the highest floods, and sediment deposited between the two sets of dikes raised the area and further confined the channel. He built more than 300 km of earthen embankments and planted nearly a million willow trees to stabilize the dikes. Now moving its huge sediment load in a confined channel, the Yellow River rapidly built out its delta and raised its bed by an extraordinary two metres in 13 years. Access to the ocean was reduced, the overwhelming volume of sediment could not be dredged, and the river even flowed upstream at times. Part of the problem was an increased sediment yield from deforestation on the Loess Plateau.

Ascending the throne in 1821 after disastrous floods, the Daoguang Emperor of the Qing Dynasty, a diligent and cautious ruler, knew that the Mandate of Heaven was faltering. But why did the cost of Yellow River regulation continue to rise, sometimes exceeding ten

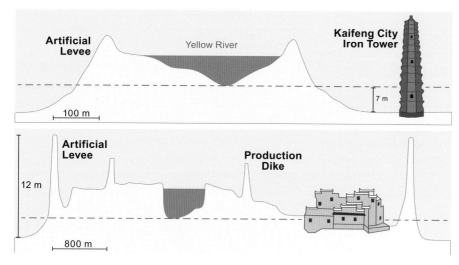

Figure 16.5 Profiles of the Yellow River at Kaifeng (above) and downstream from Kaifeng (below). In both profiles, the elevation of the riverbed is above that of the adjoining floodplain, with the river held in place by artificial levees. Levee breaks can cause widespread inundation of the North China Plain. Settlements are present in places between the artificial levees. After Chu, 2014.

per cent of state revenues? It had to mean corruption—and there were rumours of gambling and lavish parties. If a dike broke, river officials were guilty of inexcusable failure and faced demotion and exile. But as historian Randall Dodgen points out, the emperor was out of his depth in the wilful, sediment-charged river. He failed to understand that costs were spiralling because the never-ending supply of silt raised the riverbed high above the plain, requiring progressively taller dikes and sometimes several years to repair breaches. Gone were the days when armies of conscripts could be dragooned. 'Year after year,' declared the Emperor, 'military and river repair funds have been needed at the same time. No area of state finance is free of constant worry over shortages'.[14]

Catastrophic dike breaks in the 1840s

River engineer Li Yumei came from a poor family and survived a false murder charge and a corruption scandal involving his superior.[15] While inspecting a river crossing, Li heard of a threatened breach near Kaifeng and, hitching a ride on a wagon, he found that a shifting channel was eroding a crucial dike. Confident to the point of arrogance, he ignored local officials and ordered the construction of a willow dam, which blocked the dangerpoint. During a river crisis and heavy rains in 1835, with no dry earth for construction and little possibility of transporting rock from the mountains, he used bundles of brick to divert a channel nearly 400 m wide that was threatening the main dike.

The crisis was cheaply averted to the emperor's pleasure, and bricks became a standard river tool.

In 1840 while inspecting a dike, Li ate a large lunch and died later that day. Local people paid their respects as his casket passed, and the wealthy sought permission to establish temples in his name. It was a custom to encourage yellow snakes in the floodwater to climb onto platters, after which they would be taken to the Great King Temple and revered as river spirits. Encouraged in the name of many river spirits, one snake responded only when addressed as 'Great King Li'. In 1877 the state officially recognized Li Yumei as a Yellow River deity.

In August 1841, the river poured through a kilometre break near Kaifeng, which became an island in a vast lake. Considered incorruptible, Governor-General Wenchong had been selected deliberately for his lack of technical experience and a hands-off management style, but he alienated the river engineers on whom his success depended. Speedy action might have saved the day, but Wenchong had little experience, much of the plain was flooded, and roads were impassable. The river roared through the breach with a waterspout seven metres high, seemingly unstoppable. To enter the city, officials had to climb on ropes up walls that were collapsing under the relentless battering of the flood waters. Eight hundred Muslim inhabitants, under orders from their religious elders, laboured in the turbulent water to construct dikes. By mid September the situation was desperate, but officials petitioned the river gods and within a few days the main current split, taking pressure off the walls. Kaifeng had survived but was choked with silt and surrounded by a lake into which the Yellow River continued to pour.

To the emperor's envoys, Wenchong seemed dismayingly ignorant of river dynamics, justifying his failures rather than reflecting on his errors. An enraged emperor sentenced him to stand each day for three months on the riverbank, humiliated and wearing a heavy wooden collar, after which he was exiled.

River engineer Linqing was brought in to replace Wenchong. Linqing had learnt hydraulics on the job, publishing *An Illustrated Guide to River Engineering Tools*. He understood people well, and if he identified mismanagement, he did not necessarily punish the guilty but remained attentive to them. With the survivors facing a winter of starvation, he knew that only organized relief work could stave off violence, and he moved the refugees into areas where river troops could control them while he requisitioned the necessary boats, skilled watermen, and carts for transport. Reconnoitring the area below the breach, he discovered where the river was occupying old courses and creating new ones.

Repairing the breach was a staggering task. As January 1842 set in, boat crews diverted ice floes away from the construction areas. The breached earthen dike, 25 m wide at the base, was strengthened and extended for 16 km, and the breach was plugged with bundles of stalks and earth that were staked down in the boiling floodwater. Meanwhile a diversion canal more than 100 km long was dug to draw flow from the breach, ready to be cut through at a moment's notice.

By February the 'Dragon Gate' of the breach had narrowed to 12 m. The diversion canal was cut through and most of the river rushed into its old course. But waves from a violent winter storm collapsed part of dike, drowning several hundred river troops working in the darkness. Wages were doubled as feverish work proceeded by moonlight and through snowstorms. Early in March word came of an approaching flood wave, and Linqing knew that it was now or never, even though the breach was still 25 m deep. After sacrifices to the river gods, a final bundle of stalks was forced into the gap and pounded down, closing the breach.

A few months later, waves slashed a breach of 600 m through a dike downstream from Kaifeng. The river cut through the Grand Canal and carved a new course to the Yellow Sea, flowing as a sheet of muddy water 50 km wide. A displeased emperor removed Linqing from office. But there was more trouble to come. In the summer of 1843, a kilometre-wide breach opened in the dike at Zhongmou where, as an official noted, 'The people work on the river, are impoverished by the river, then die in the river'.[16] The local governor rushed to the site but could only stabilize the ruptured dike ends, and he too was removed from his post. Ten thousand refugees crowded on the dikes and Linqing, considered capable of self-renewal by the emperor, was given charge of repair materials, which were brought by road to the lake, loaded onto boats, and reloaded onto carts for transport to the dike.

As winter blizzards set in, an army of rebellious labourers dug a diversion canal and repaired the dike, under guard by river troops. The frozen bodies of emaciated workers lay on the dikes where they fell. Linqing arranged prizes for exceptional workers and he awarded boots and hats to the river troops. Many boatmen drowned when ice floes capsized working vessels, and lotus lanterns were hung to assist the drowned souls to the next world. As the breach narrowed, the channel deepened to 45 m. Finally, in February 1845 a stalk-and-earth plug was rammed down and the breach closed. The governor was reinstated and Linqinq promoted, although not to his original status.

Floods in the succeeding years provoked rebellions, and sea routes began to replace the Grand Canal. In June 1855, a breakout downstream from Kaifeng forced the Yellow River into a course 300 km to the north, and there was no alternative for Daiguong's successor but to let the river go. Floods over the next decades devastated the farmlands and took the lives of millions, entombing houses in silt up to the roofs and floating away coffins exhumed from shallow graves.[17]

Exhausted and in poor health, Linqing died in 1846. Not long before his death, he wrote:

I stood still for a time as the night gradually deepened. Then I closed the window, raised the wick of the lamp and continued reading … I recalled that when I was young and my ambition was still unbounded, seeing that all so-called worldly affairs and accomplishments were achieved through action, I questioned the place of these two words 'tranquillity' and 'peace'. Now, after more than thirty years of official service, I finally understand.[18]

Farming increases the sediment load

Agriculture augmented the enormous sediment load of China's rivers. The Yellow River landscape began to change more than 8000 years ago when Neolithic inhabitants cultivated millet and eventually rice, using fire to clear vegetation.[19] As the forests were cut down and the soil eroded, sediment clogged the valleys. Climatic changes that included a waning monsoon and the El Niño rhythm also affected the river, and there were decades of unusually cold winters when the Yellow flooded powerfully and froze repeatedly.[20] By the end of the first millennium BCE, human and natural effects were so intertwined across the Loess Plateau that archaeologists are commonly unable to distinguish them.

In southwest China, rice cultivation over millennia and deforestation for salt works and mines strained the carrying capacity of the land.[21] As the eighteenth century dawned, the burgeoning population settled on the valley sides, and an official noted,

> The mountains whose sides had been ploughed collapsed in their entirety. An immeasurable amount of water laden with sediments and massive boulders several fathoms in size … filled the channel … so that not a drop of water flowed through it.[22]

Tens of thousands of men dredged and embanked the Miju River each year. Officials took their work seriously and an 1845 record noted 'People's survival is not to be treated as a children's game. Nor should crises be made into a source of profiteering'.[23] For the workers it was back-breaking toil, as poet Gao Guizhi recorded:

> I've been clearing the river out. Then, a second time, getting it dredged.
>
> Its compacted muds were coming apart, with numerous tear-filled runnels.
>
> Though up on the heights now, for Autumn Day, there's no glimpse of a hopeful perspective,
>
> My muscles, and scaffolding of bones, having dissolved into nothing.[24]

Environmental warfare on the Yellow River

The use of rivers in war was frequent in China, as in ancient Mesopotamia, and in 651 BCE some states had signed a treaty banning levee breaches in conflicts.[25] Seeking to delay the advance of the Japanese army in June 1938, the Chinese nationalist government breached the southern dike of the Yellow River in flood at Huayuankou in the largest act of environmental warfare in history, leaving the downstream channel almost dry.[26] With little warning, villagers were overwhelmed and an estimated 844,000 died from drowning, disease and famine, with millions more displaced. Covered with silt and with irrigation channels destroyed, much of the land was left uncultivable. The Japanese army suffered only minor losses.

Breaching the dike came to symbolize the nationalist government's indifference to its people, and the area became a fertile recruiting ground for the communists. By 1947, the river had been forced back to its course with engineering assistance from the United Nations. Two years later, Mao Zedong came to power and the People's Republic of China was established.

Floods on the North China Plain

In 1933, the Yellow River was out in force and missionary Andrew Thomson travelled by boat with officials through the drowned sorghum crops to assess the damage.[27] They found villages reduced to islands with collapsed mud houses, as well as piles of furniture, pottery, and spinning wheels. Some villagers had clung to the treetops for days until the floods abated, one man supporting his elderly mother. The floods had destroyed 4000 villages with a population of three million people. Destitute refugees crowded on trains, heading out of the region to nowhere.

Arriving on the North China Plain in 1906, Andrew and Margaret Thomson developed a deep respect for the sturdy villagers with their simple dignity and sense of humour. Thomson worked with local magistrates to bring flood relief, and he joined city officials in negotiating with warlords whose armies were converging on the region. In the aftermath of the 1933 floods, Thomson used his own money to build a hospital for flood refugees, staffed by dedicated Chinese health professionals. After further floods, he used the family's life savings to purchase rail cars of grain, which he sold at interest to local businessmen, funding construction work that kept an estimated 11,000 villagers from destitution. A good harvest the following summer recouped the family savings.

Staying on in war-torn China against consular advice, Andrew's daughter Betty and my uncle Godfrey Gale, medical workers at a hospital on the Yellow River, were placed under house arrest by Japanese authorities.[28] They buried archaeological relics at the medical school and milked the goat of a Chinese friend through the perimeter fence to provide for the hospital. (But there was a human side to the invasion: Japanese soldiers, missing their families, would visit the children's ward and sit holding the hands of small children). They travelled to a civilian internment camp in a barge up the Grand Canal (a voyage that Godfrey had once urged Betty to take), sheltering under a plank from the pouring rain.

River poets madly singing in the mountains

Poets Du Fu and Li Bai, often found madly singing in the mountains,[29] were caught up in a rebellion in 755 CE, during which Chang'an was overrun. Du Fu recorded conscripted men marching silently to their doom, and his cousin was killed, 'on your face three years of dust'. Meanwhile, the rivers flowed steadily on:

> A man who has feelings must weep upon his breast,
>
> But you, river waters, river flowers, do you never care?[30]

Jailed for several months after the rebellion, court poet Li Bai made the dangerous passage to exile through the Three Gorges on the Yangtze:

> Azure heaven pinched between Wu Mountains,
>
> riverwater keeps streaming down like this,
>
> and with riverwater cascading so suddenly
>
> away, we'll never reach that azure heaven.
>
> Three mornings we start up Huang-niu Gorge,
>
> and three nights find we've gone nowhere.
>
> Three mornings and three nights: for once
>
> I've forgotten my hair turning white as silk.[31]

A wanderer for much of his life, Li Bai died as he had lived with rivers. He is reputed to have toppled from a boat in a drunken state and drowned in the Yangtze while trying to embrace the moon.

PART 5: ENGINEERED RIVERS

CHAPTER 17

DEAD AND WOUNDED RIVERS

A dead river

The Sebeş River enters Transylvania through the Carpathian Mountains near Dracula's castle. Invading hordes repeatedly occupied the river valley, but today the Canadian geology students are far from the invading horde that they might otherwise be. Most have succumbed to a short-lived infection that spread through the bus, and there is an occasional rush for a sheltering bush. Our geological instructor is reading my tourist guide to Romania with some annoyance, able—like most Transylvanians—to recite historical chapter and verse. 'They think this happened in 1293?' he roars, startling the driver.

In a dry summer, the Sebeş River still manages a little flow on its way to the Danube. As we lunch by the stream, harness bells ring from carts laden with hay, in one of which an old man sleeps as his nag trots home. I feel close to the heart of an ancient river, flowing much as it has done since the Carpathians rose.

The bus roars up the road. Around the next bend, we are confronted by a high, curved concrete wall where an elderly guard sits incongruously on a stool. I snap a few shots: the impressive dam wall, a trickle of water in the channel below (Fig. 17.1). Three more dams block the river before we cross the pass, diverting the water into pipes that emerge at turbines downstream. Tourists enjoy the view of the reservoir, and a serious little girl sells ice-cream from a booth. But below the dam the channel is dry.

Rural Transylvania has deluded me into thinking that this is a pristine river. American conservationist Aldo Leopold considered the Carpathians to be the last European wilderness,[1] but the river is divided by a staircase of dams into reservoirs and trickles, or it has no flow at all. The Sebeş is a *dead river*, a term coined by John Muir more than a century before.

Figure 17.1 Dam on the Sebeş River, Romania.

John Muir and the rivers of California

In 1867 John Muir, environmentalist, writer, and co-founder of the Sierra Club, set out 'joyful and free' to walk from Indiana to the Gulf of Mexico.[2] His foolhardy plan was to take ship to South America and float down the Amazon.

Muir's family had emigrated to Wisconsin in 1849, the same year that a wave of gold diggers rushed to California. The young Muir was intoxicated by the New World, but he was soon disillusioned by the sheer violence with which the settlers slashed and burned the magnificent forests. It was a far cry from the *Peaceable Kingdom* painted by early American artist Edward Hicks, depicting harmony between the continent's plants, animals, and humans.

An experienced millwright, Muir moved to Indianapolis. There a file flew into his right eye, leaving him sightless and in suicidal despair when his left eye went into sympathetic shock. Within a month, however, his sight recovered. Abandoning industrial work, he headed south carrying a small pack that included a compass, a case for his plant collection, and three books: a New Testament, the poems of Robbie Burns, and Milton's *Paradise Lost*.

In the Appalachians, Muir stayed with a blacksmith, a godly man who suggested that collecting plants was a waste of time. Muir replied that King Solomon had studied the plants growing from Jerusalem's walls. The blacksmith, impressed by his biblical knowledge, urged him to abandon his dangerous journey. It was a serious warning. On a lonely path, he came face-to-face with a mounted guerilla band but ducked past them with a cheery 'Howdy', taken for an itinerant herbalist. In Georgia, burnt fences and 'woods ruthlessly slaughtered' testified to the recent war. Accosted in Florida, he thrust his hand into his (empty) pistol pocket, and snarled, 'I allow people to find out if I am armed or not'.[3]

Muir was intoxicated by the mountain rivers that he forded or swam, but he passed scores of dams and water mills built by settlers who were fast changing the streams. At Savannah he slept in a decorated graveyard, an 'art blunder' that Nature was working to erase. Anxious about the Florida alligators, he nevertheless considered them beautiful in God's eyes and 'honorable representatives of the great saurians of an older creation ... blessed now and then with a mouthful of terror-stricken man by way of dainty!'[4]

Muir eventually reached California and found employment as a shepherd. He grew to dislike the 'hoofed locusts' that destroyed the wildflower meadows, tolerating their occasional loss to those superior beings the wolves. In the Yosemite Valley, he dammed a creek and dug a mill race to a waterwheel, once narrowly escaping death when a shift in the wind brought a thousand-foot waterfall down on his head. He found glaciers buried under mud and used stakes to measure their flow rates, impressing Louis Agassiz. Evolution bonded him to all creatures in the wild where 'I will touch naked God'.[5]

Muir Woods near San Francisco was established in 1908 as a preserve for the gigantic redwoods, the plight of which had caused Muir to declare, 'No doubt these trees would make good lumber after passing through a sawmill, as George Washington after passing through the hands of a French cook would have made good food.'[6] But the 'forty-niners', named for the year in which most of them reached the goldfields, devastated California's rivers in the search for gold. Muir wrote in 1875 that he had seen a *dead river* near Sacramento, something worth travelling round the world to see.[7] Then Muir learnt that plans were afoot to dam the Hetch Hetchy Valley near Yosemite to supply San Francisco with water. 'Dam Hetch-Hetchy!' he exclaimed. 'As well dam for water-tanks the people's cathedrals and churches, for no holier temple has ever been consecrated by the hand of man'.[8] He castigated the developers who claimed that a beautiful reservoir would enhance nature: 'These temple destroyers, devotees of ravaging commercialism, seem to have a perfect contempt for Nature, and, instead of lifting their eyes to the God of the mountains, lift them to the Almighty Dollar.' The dam was approved in 1913 a year before Muir's death, the vanguard of an engineering revolution that, within a few decades, changed the world's river landscapes for ever.

Hydraulic mining in California and the rise of big dams

Rivers around the world had been appropriated for human use since the Neolithic Revolution.[9] Along European rivers, watermills became increasingly important during the first millennium CE (Fig. 17.2). In England,

Figure 17.2 Water mill, weir, and millpond on a river in France. Shutterstock by JP Wallet.

the 1086 Domesday Book commissioned by William the Conqueror listed a watermill for every 250 people. From the Middle Ages to the seventeenth century, salmon declined by 90 per cent across northwest Europe as streams were dammed for millponds, covering the gravel with mud and preventing the fish from reaching their upstream spawning grounds.

But it was in North America that rivers were most fully harnessed. By 1840 more than 65,000 watermills with dams several metres high were operating in the northeastern United States alone, each ponding the river for some kilometres upstream.[10] So intense were the changes that, in little more than a century, all knowledge of the original rivers was lost. Thick coal seams at Pittsburgh in the Ohio Valley astonished Charles Lyell when he visited the area in 1846, and the smoke that enveloped the city to the tops of the church towers caused novelist Anthony Trollope to consider Pittsburgh the blackest place he had ever seen.[11] Barges laden with coal drifted down the wild Lehigh River, many breaking up in the rapids, and canals changed

river courses during their fleeting lifespan, clogging them with coal dust.

Then in 1848 gold was discovered in California. Hundreds of thousands of would-be miners converged on San Francisco, and as yields from gold panning declined along the Sacramento River system, prospectors found gold in older terrace gravels on the hillsides. The 1859 discovery of silver in the Comstock Lode of Nevada opened further prospects. Infected with mineral fever, Mark Twain blasted shafts until his funds ran out, later working ineffectively with a shovel in a quartz mill. In his view, 1848 replaced the desire in America to earn a livelihood with a lust for wealth.[12]

Gold mining from river sediment required water to feed sluices for sorting the gold.[13] Using reservoirs behind wooden dams in the mountains, California miners developed *hydraulic mining* in which high-pressure water cannons or monitors blasted away the sediment cliffs (Fig. 17.3). The dams allowed gold to be extracted year-round and far from the rivers, and by 1879 more than 1500 km of ditches and sluices were

Figure 17.3 Hydraulic mining in the Sierra Nevada of California in the 1880s, using high-pressure water cannons to wash alluvium for gold. Photo by Carleton E. Watkins, USGS.

feeding the monitors. Large mining operations ran through the night by the light of blazing fires.

Hydraulic mining spelled disaster for the landscape. Rivers were choked with debris and became dangerously shallow, flooding the rich agricultural land with tailings. After disastrous floods, the US Army Corps of Engineers confined the rivers with artificial levees to deepen the channels and hasten sediment transport, much as Pan Jixun had done for the Yellow River. California realized that its future lay in agriculture, not mining, and in 1884 a federal court halted the dumping of tailings in Californian rivers.

Twenty years later with gold prices high, companies sought a renewal of mining. Called in to make an evaluation, veteran geologist G.K. Gilbert found that valleys near Sacramento were choked with some 60 m of gravel and that forests had been buried where they stood.[14] A debris wave was moving downstream to San Francisco

Bay, where dredging had already removed more material than had been extracted during construction of the Panama Canal. In a remarkable study, Gilbert surveyed the abandoned mine workings, calculating from their volume that more than a cubic kilometre of tailings had been released, more than the volume of ash erupted from Mount St Helens in 1980. Decades of river incision through the young alluvium had left flights of terraces, normally formed over hundreds of thousands of years. Every river process had been speeded up.

Faced with economic disaster, the California mining companies diversified into suppliers of water and electricity, using the dams to run turbines for power generation. The first hydroelectric plant had opened in Wisconsin in 1882 as a run-of-the-river plant without a reservoir, but high-head dams could generate much more power, and the development of alternating current allowed efficient power transmission. By 1895 electricity from the Folsom Dam on a Sacramento tributary was powering San Francisco's streetcars,[15] and by 1900 more than 40 per cent of America's electricity was from hydropower.

The completion in 1936 of the iconic Hoover Dam on the Colorado River marked the real watershed (Fig. 17.4). As far back as 1905, corporate farms and the city of Los Angeles had pressed for a dam after the river flooded the Imperial Valley, overwhelming a canal and creating the Salton Sea.[16] Constructed by the Bureau of Reclamation (BuRec), the Hoover Dam was a formidable undertaking with its massive concrete face, rock-cut diversion tunnels, and Art Deco facings. As the Depression set in, thousands of labourers earned a dangerous living at the dam, but the work finished ahead of schedule and Reservoir Mead filled behind the dam.

Hoover Dam transformed BuRec into a powerful engineering agency. From the 1930s onwards, BuRec and the US Army Corps of Engineers built big dams for the Tennessee Valley Project as part of Roosevelt's New Deal, as well as a string of silt-trapping dams on the Missouri. The Grand Coulee Dam blocked the world's largest salmon run on the Columbia River, and a cascade of dams and reservoirs provided navigation on the Snake River. The rivers changed from *floodplain rivers* to *reservoir rivers*. Although hydropower is an important energy source, in the words of writer Marc Reisner, BuRec gave America's free-flowing rivers a blanket death sentence.[17]

Dams as a cure for awful nature

Fast-forward 80 years from the opening of the Hoover Dam.[18] The Earth's rivers are now home to more than 58,000 large dams, defined as taller than 15 m (equivalent to a three-storey building), with some more than 300 m high. Of these, more than 23,000 are in China, 9200 are in the United States, and some 900 in Canada. Considering dams of all sizes, the United States has more than two million. By 2007, only 21 of the world's longest 177 rivers were running freely from their headwaters to the ocean.

There are nearly 17 million reservoirs of all sizes on Earth. Their volume is equivalent to lakes Michigan and Huron combined and about four times the volume of all the world's rivers. Evaporation from the reservoirs is

Figure 17.4 Hoover Dam, completed in 1936, and Reservoir Mead on the Colorado River. Note the 'yellow belt' on the rocks upstream of the dam due to varying water levels. Shutterstock©scullydion.

about five per cent of global river flow. Their huge mass is thought to have caused a slight change in the Earth's orbit.[19]

On the positive side, hydropower generates nearly 20 per cent of world electricity. Dams also provide irrigation, municipal water, flood control, navigation, industrial-waste storage, and recreation. On the negative side, tens of millions of people have lost their land, and hundreds of millions have been affected downstream to the river mouths. Sickness from water-borne parasites and skin diseases has proliferated. As drowned vegetation decays, many reservoirs have become 'methane factories', and some reservoirs and generating stations produce as much greenhouse gas as a coal-fired thermal station.[20] Hydropower is not 'green energy', and it is cheap only because water is considered cost-free.

For rivers, dams are an unmitigated disaster. As many as 20,000 freshwater species are at severe risk or already extinct, with much of the loss linked to dams that have regulated river flow and fragmented aquatic populations, as well as destroying habitat and trapping fertile sediment and nutrients.[21] Few large dams allow migrating fish to pass. A hydrodam construction boom is underway on the Amazon, Mekong and Congo systems—the world's river basins with the highest biodiversity—without basinwide planning or consideration of the cumulative consequences.[22]

Dams are magnificent engineering achievements, measuring people against nature like an engineering Olympic Games.[23] Soviet engineers saw themselves as making 'mad rivers sane', and a Canadian engineer commented, 'In my view, nature is awful and what we do is cure it'.[24] Hoover Dam exemplified greatness and domination, and similar terms apply to dams around the world. In the words of dam expert Patrick McCully in his book *Silenced Rivers*,

> Politicians and developers have for most of this century expounded that a river has no value unless it is in some way *controlled* (and not just used) by humans. This belief negates the intrinsic worth of rivers—the veins of the hydrological cycle, shapers of the landscape, and providers of life to many of the earth's species; it negates their cultural, aesthetic and spiritual importance, and it negates the economic value of unregulated

rivers to the hundreds of millions of people who depend on them for drinking water, food, transport, recreation and other uses. The wasted river ideologues are justifying not human use of rivers, but the expropriation of rivers from one set of users to another.[25]

Taming China's river dragons

The Communist forces under Mao Zedong and Zhou Enlai regrouped near the Yellow River after the Long March. Here in 1939 Xian Xinghai composed the *Yellow River Cantata*, set to a text by Guang Weiran. Guang made crossing the dangerous river a metaphor for the struggle against Japanese occupation:

> Have you crossed the Yellow River?
> Do you remember scenes of the boatmen
> risking their lives to battle the perilous waves? …
> Do not fear the mountainous waves!
> Boating on the Yellow River is like fighting at the front,
> Unite and forge ahead![26]

The libretto continues 'Within our hearts, enmity and hatred swirl like the rapids of the Yellow River!', and the cantata ends with a call to defend the river and China. Mao attended the second performance, which remains a celebrated musical work.

In 1958, the *Sanmenxia Cantata* by Guang Weiran was set to music by Xi Xianhe. The cantata celebrated the construction of the Sanmenxia Dam, designed to trap sediment from the Loess Plateau and leave the Yellow River 'clear as jade'.[27] The cantata's symbolism, however, was markedly different. The dam is like a battlefield, proclaims the text, and the river will not be allowed to trouble the people: emperors and people failed to subdue the water but revolutionary anger is more powerful than the river. Once people hated the river but now they love it, and the dragon is laughing and singing. Today there are thousands of heroes like the legendary Yü the Great. 'The old world must be changed!'[28]

Huang Wanli was an engineer and poet who had visited many big US dams.[29] After returning to China in 1937, he surveyed on foot 3000 km of Chinese rivers. Predicting that the Sanmenxia reservoir would fill rapidly with sediment, Huang stated that the goal of a clear Yellow River distorted the laws of nature and that sediment should be funnelled through sluice gates low in the dam wall, rather than being trapped. Unwavering in presenting the truth as he saw it, he published an allegorical story that was all too readily interpreted by a furious Mao, and he was sentenced to hard labour at Sanmenxia, studying siltation problems at night after the day's work. After Mao's death he was rehabilitated, and his poems were distributed to colleagues on his 80th birthday. Premier Zhou Enlai also had reservations. Finding only plans for a high dam wall and storage of water and sediment, he chided the planners for their extravagance of scale. 'Some mistakes can be remedied in a day or a year,' he fumed, 'but mistakes in the fields of water conservancy and forestry cannot be reversed for years.'[30]

By 1962 Sanmenxia was generating hydropower behind a high dam wall, and more than 400,000 people had been relocated. Within three years, five billion tons of sediment had accumulated behind the dam and the turbines had to be removed. Soon, the sediment had backed up almost to Xi'an, and a worried Mao told Zhou Enlai to bomb the dam with aircraft if nothing else worked. Eventually low-level sluice gates were opened at great expense, and turbines were installed with a reduced capacity. By 1999 the Xiaolangdi Dam downstream from Sanmenxia was in place, and within eight years, a delta had advanced for tens of kilometres into the Xiaolangdi reservoir.[31] The two dams and others upstream had trapped some 19 billion tons of silt within a few decades, equivalent to the weight of 50 Empire State buildings. For some days each year since 2002, the world's largest hydrological experiment has released a raging flood of turbid water and silt from the reservoirs, eroding the channels and extending the delta at the Yellow Sea.[32]

Irrigation takeout has been largely responsible for the unprecedented drying-up of the Yellow River since 1972. The main channel had ceased to flow only ten times in nearly four millennia, but in 1995 the river was dry for nearly 700 km. Two years later, no-flow days exceeded 200. The Yellow River is now an *anthropogenic seasonal* river,[33] a far cry from the free-flowing torrent that Li Bai celebrated.

River megaprojects have continued to be debated across China. Coming into full operation in 2012, the Three Gorges Dam on the Yangtze is nearly 200 m high and provides hydropower and flood control, as well as navigation along a reservoir 600 km long (Fig. 17.5).[34] Between one and two million people were displaced by the reservoir, probably more than for any other project worldwide, and the barren 'yellow belt' between low and high reservoir levels covers an enormous 350 square

Figure 17.5 Three Gorges Dam on the Yangtze River, 2.3 km long. The reservoir extends for 600 km upstream of the dam. Shutterstock©isabelkendzior.

kilometres. The Yangtze River dolphin and the Yangtze paddlefish, seven metres long, are probably now extinct; the latter was formerly the longest freshwater fish with an ancestry that stretched back to the Jurassic.[35] The South–North Diversion Project transfers water for more than 1000 km from the Yangtze to water-strapped north China—the largest engineering project in China since the Grand Canal. According to critics, measures that include supplying water only as needed and recycling wastewater could have rendered this project unnecessary.[36]

While recognizing the importance of energy generation, journalist Dai Qing commented in 1998 about the Three Gorges Dam:

> I think that 90 per cent of the Chinese people are really opposed to the project. The only ones in favour of it are those politicians and people in the hydroelectric departments who profit in some way—either financially, or through prestige, glory, and promotions. They are after power at the expense of the country … If the Three Gorges could speak, they would plead for mercy.[37]

An indigo river in India

The industrial town of Pali, famous for its cotton mills, stands on a tributary of the Luni River in the Thar Desert of India. Now in February the broad river is a trickle and the monsoon rains are a distant memory. Laid out on the gravel to dry are strips of newly dyed cloth, twisted where the wind has tugged them. On a yellow cloth sits a group of women with strong, characterful faces, amused at the unexpected appearance of geologists. Presently they walk back to town with bundles of cloth on their heads. An oxcart laden with cloth trundles across the bridge, the driver perched on the load.

Today is blue day at the Pali dying works. From under the bridge comes a plume of indigo water with a powerful chemical stench, washing up in steaming pools all down the river (Fig. 17.6). A black-winged stilt with its long red legs pecks at insects in a stinking pool of blue water.

Some rivers have been cleaned up. Victorian sewage systems rejuvenated the Thames in London (see Chapter 21), and sawdust from lumber mills, which had drawn criticism from Oscar Wilde during his visit, no longer clogs the Ottawa River in the Canadian

Figure 17.6 Bandi River at Pali in northwest India, a tributary of the Luni River. The water is stained blue with dye from a textile works.

capital. No one has a right to pollute water, he stated, a common inheritance that should be left unpolluted to our children.[38]

But industrial effluent continues to devastate rivers around the world.[39] Millions of cubic metres of toxic waste and cyanide have roared down rivers after the collapse of tailings dams, damaging among others the Danube, the Guadalquivir in Spain, and the Fraser in Canada, the latter disaster releasing 25 million cubic metres of tailings and water. In 2019, a tailings-dam collapse at Brumadinho in Brazil released a debris flow of at least 12 million cubic metres, laced with iron-rich tailings and toxic materials, into the Paraopeba River, a tributary of the São Francisco (Fig. 17.7). At least 250 people were killed and tailings materials spread down the river for at least 240 km, increasing mercury levels in the water to 21 times the acceptable level.

Hydrocarbon, industrial, and agricultural contamination has been widespread. Polluted by tarsand extraction, the Athabasca River of Canada has long exceeded toxicity guidelines, contrary to industry and government claims. In 2010, a pipeline rupture released more than 3000 cubic metres of crude oil into the Kalamazoo River system in the United States, and 16,000 dead pigs floated down a river near Shanghai in 2013. Burning oil slicks on the Cuyahoga River in Cleveland have engulfed workers in walls of fire, and an oil slick burned on the Hudson River in Manhattan in 1921.

Following the 1986 Chernobyl nuclear disaster, a radioactive plume reached the Black Sea from the Dnieper River system in Ukraine, the third largest

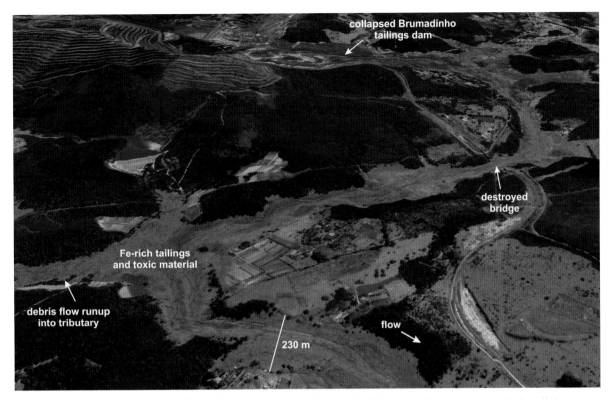

Figure 17.7 Brumadinho iron tailings dam collapse into Paraopeba River, near Belo Horizonte in Brazil, January 2019. The river was swept by a debris flow laced with iron tailings and toxic material. The field of view shown is 3 km in a straight line from the collapse site downflow. Maps data: Google, Maxar Technologies.

river in Europe after the Volga and Danube, with a drainage basin of half a million square kilometres. An important source of water for irrigation and human use, the Dnieper was heavily contaminated from the initial fallout, and runoff from contaminated floodplains and nuclides attached to suspended sediment have continued to boost the river's radioactivity. It may take 40 years for the Dnieper to return to its pre-disaster condition. In 2011, a radioactive plume moved down the Abukuma River in Japan after an earthquake damaged the Fukushima Nuclear Reactor. Radioactive particles from the nearby uplands were rapidly transmitted to the river system by water flowing through terraced ricefields.

In the past three centuries, humans have changed rivers worldwide more dramatically than anything in the previous four billion years. In 2007 the World Wildlife Fund listed the Top Ten Rivers at Risk from a variety of causes—Ganga, Yangtze, Salween, Mekong, Indus, Danube, La Plata, Rio Grande, Nile, and Murray-Darling.[40] We live among wounded, crippled, and dead rivers everywhere.

CHAPTER 18

COLLAPSING AND CLOSING DAMS

Collapsing dams

High in the mountains above Denver, where the Olympus Dam impounds the Falls River, an enormous spread of boulders emerges from a side valley (Fig. 18.1). This is the Roaring River Fan, created instantaneously in 1982 when deteriorated caulking in a discharge pipe breached the Lawn Lake Dam. In less than a minute more than 100,000 cubic metres of water exploded down the Roaring River and into the Falls River below, killing three people and demolishing a second dam before the rushing water, heading for the Colorado plains, was contained by the Olympus Dam. The flood cut down more than 20 m into the Roaring River valley, and 750,000 tons of boulders were set in motion.[1] At the toe of the fan, tree trunks and car-sized boulders jammed into the forest.

Another dramatic dam collapse was the 1976 failure of BuRec's Teton Dam on the Snake system in Idaho, which killed eleven people.[2] The earthen dam had been anchored in volcanic rock so cavernous and fractured that water pumped down test boreholes vanished into the ground. Technical objections, however, were swept aside. As it filled, the dam collapsed and a fifteen-metre wall of water rushed down the valley across a three-kilometre front at nearly 30,000 cubic metres a second. A battering ram of mud and debris smashed into the town of Rexburg and a gasoline storage tank ignited, lacing the flood wave with flaming petroleum. The flood stripped the soil from more agricultural land than the dam would have irrigated.

Figure 18.1 Boulder fan of the Roaring River, Colorado, generated by collapse of the Lawn Lake Dam in 1982.

But the worst dam disaster of all remains barely documented.[3] In 1975 a typhoon laid down a metre of rain in three days across central China, with such force as to kill birds in flight. The runoff was far beyond the design capacity of a series of dams in the Huai River system. Water poured over a dam more than 100 m high as workers struggled to repair the failing embankment in the darkness. Someone shouted: 'The river dragon has come!' and the dam collapsed. A flood with more than five thousand times the volume of the Lawn Lake flood roared down the valley at nearly 50 km an hour, destroying 62 dams. More than 85,000 people drowned in a wall of water that subsided into a lake 300 km long, and millions were left clinging to trees and rooftops as famine and epidemics decimated the survivors. Altogether the death toll may have been as high as 230,000.

By 1981, China had formally acknowledged 3200 dam collapses, nearly four per cent of the country's dams.[4] Years later during the Three Gorges Dam debate, critics countered unsuccessfully that large dams should not be the lynchpin of flood control, as the Huai catastrophe had shown. Had the world outside China known of the dam collapses, Patrick McCully suggests, the disaster would have done for the public image of dams what the 1986 Chernobyl disaster did for nuclear power.[5]

The High Aswan Dam drowns the Nile

I was ten years old when I started to correspond with an elderly lady in Kuwait. Her letters brought an exotic flavour of life in faraway places. They also brought brightly coloured stamps showing the gigantic statues of Pharaoh Ramses II at the temples of Abu Simbel. The statues had been carved more than three thousand years before in the solid sandstone by the First Cataract of the Nile, and passageways guided the first rays of the rising sun into the sanctuary. Soon, however, the statues would drown below the rising waters behind the High Aswan Dam. UNESCO launched an international appeal to save the statues which, in an extraordinary engineering feat, were cut apart block by block and reassembled at a higher level (Fig. 18.2). The stamps found their way into my stamp album, and the dam and the drowned temples into my imagination.

Hassan Dafalla became Sudan's District Administrator in 1958 for the region that later included the

Figure 18.2 Statue of Pharaoh Ramses II at Abu Simbel, Egypt, under reconstruction at a higher level in 1967 before the Aswan High Dam flooded the temple. Photo by Per-Olow Anderson.

reservoir.[6] Flying into Wadi Halfa near the Egyptian border, he was impressed by rows of date palms, considered a gift from heaven by the Nubian population to compensate for the confinement of the river valley. Soon he began to understand how profoundly Nubian tradition centred on the Nile. Seven days after the birth of a child, local women carried the infant to the river where a meal was eaten and the leftovers thrown into the water, along with the child's clothes, to 'feed the angels'. The women carried river water back to the mother, who swirled a mouthful over the child saying 'My child is safe'. Within the narrow valley, crops flourished on the rich river loam following recession agriculture. As novelist Anne Michaels recorded in *The Winter Vault*,

> The silt, like the river water, also had its own unique intimacies, a chemical wisdom that had been refining itself for millennia ... the Nile silt was like flesh, it held not only a history but a heredity. Like a species, it would never again be known on this earth.[7]

A year after Dafalla's arrival, agreement was reached to build the dam. Encouraged by the proposed benefits, world governments disregarded what Dafalla called the 'evil aspects' for Nubia: Wadi Halfa and tens of villages would be drowned and more than 100 thousand Sudanese and Egyptian Nubians would be dispossessed. People rushed into the streets overwhelmed by their uncertain future and envying their ancestors at rest in the graveyards. Poems of tribute to the land were set to music. For centuries they had adapted to the life-giving desert river and had taken pride in their ancestors' contribution to human civilization. Now death would come from the blessed waters of the Nile itself.

In his book *The Nubian Exodus*, Dafalla documented the resettlement plans that focused on the Atbara River to the south. Tens of thousands of livestock and birds would need transportation along with their owners. And compensation would have to be paid for the date palms. As the dam neared completion, the Nubians boarded trains decorated with the branches of the palms that the reservoir would soon drown. Concerned about the long journey, Dafalla arranged for a birthing car with hospital beds, and he handed a parcel of shrouds to the conductor in case of need. As the final whistle sounded the passengers burst into tears and a tumult of shouts resounded. At one village Dafalla was dismayed to find the station deserted, but he discovered the people engaged in a religious dance in the graveyard, decorating the tombs with branches and reading poems. Soon, the new inhabitants were coping with sandstorms and waterlogged plains along the Atbara. The move required a complete change in their agricultural practices and, as Dafalla put it, 'their ecology was shaken to the roots'.

With the Aswan Dam now sealed, Dafalla watched a torrent of water pour through Wadi Halfa, amid the roar of collapsing masonry. Rats emerged from holes carrying their young in their teeth, and the water swept along scorpions and reptiles. Soon only the tops of the palm trees and the minaret of the mosque were visible. Before the post office closed, he purchased postage stamps and stuck them on envelopes marked with the town's final date. The river had drowned.

As for Abu Simbel, it could no longer be considered a temple, its holiness seeping away as the rock saws cut it apart. 'The Great Temple had been carved out of the very light of the river,' wrote Anne Michaels, 'carved out of a profound belief in eternity … The stone had

been alive to the carvers.'[8] In the end, the real lie was not moving the temple but moving the river.

The High Aswan Dam has been a boon and a bane.[9] The dam has prevented extreme floods and generated hydropower, as well as bringing much of the Nile Delta under irrigation. But there were critics in Dafalla's time who warned that the dam would damage the delta's fertility by trapping the rich Nile silt from the Ethiopian highlands. As they predicted, the river no longer has a functioning delta because too little sediment is available to counteract sea-level rise, the coastline is eroding back, and fertilizers must replace the nutrients lost with the silt, badly affecting offshore fisheries. Large volumes of water have percolated from the reservoir into the porous Nubian Sandstone below, causing earthquakes. With a desert reservoir losing water to evaporation, Egypt—a country uniquely dependent on a single river—has less water than before.

The building of Kariba Dam, and Operation Noah

The Zambezi River runs through a deep gorge at Kariba downstream from Victoria Falls, and in 1955 work began on the Kariba Dam, at that time the world's largest human construction.[10] A coffer dam was pumped out to expose the gigantic boulders on the Zambezi's bed for the first time in thousands of years, and men working below water level could hear the boulders grinding as the swollen river swept past. They were fortunate to survive when the coffer dam was breached. With little reliable river data, the engineers had a deceptively benign view of the Zambezi and, had it not been for this exceptional flood, an inadequate spillway would have compromised the dam foundation.

As the dam sealed off the river, the Zambezi downstream became a muddy trickle where fish struggled in stagnant ponds. The workforce, many from Italy, were proud of damming such a wild river. Nevertheless, some expressed regret. 'I've never known a river like this,' said one worker. 'Of course, I'm glad we have done what we set out to do. But somehow…'[11]

An unforeseen catastrophe now unfolded. The Kariba Gorge was a prolific wildlife haunt and many creatures had no way out. Into the branches of the trees scampered thousands of monkeys. Able to withstand average river floods, they could not survive a hundred-metre rise in water level, and they clung to the topmost branches until they were swept away. Visitors in boats, filled with pity, shot as many as they could until their ammunition ran out. Local conservation officers rescued thousands of bucks, baboons, and warthogs marooned on temporary islands or drifting on logs, in what became known as *Operation Noah* (Fig. 18.3). But only a few thousand animals could be saved from what has been described as 'the greatest environmental upset ever to befall a population of animals and birds within the African continent, in the memory of man'.

Since Operation Noah, visitors and writers have grown to accept the attractive landscape of the world's largest reservoir, but the former richness of the gorge has passed out of knowledge. As human ecologist David Hughes expressed it, 'Euro-Africans yearned for water, glorified the lake and forgave the dam.' 'No single project before or since has snuffed out this much life this fast,' he notes. 'Yet, the lethal wall of concrete no longer

causes onlookers—even romantic ones—to shudder. If engineers tamed the river, writers tamed the dam.'[12]

Glen Canyon Dam: John Wesley Powell turns in his grave

The narrow concrete plug of Arizona's Glen Canyon Dam is rammed into the Colorado River canyon, spanned by a steel bridge anchored into the rock (Fig. 18.4). In the canyon walls, the Jurassic Navajo Sandstone displays red cross-beds that were once the sand dunes of Pangea. From the power stations pylons stride across the land like giants in harness, hauling electricity to the cities.

Reservoir Powell, almost 300 km long and in places 200 m deep, laps gently against the dam. The reservoir fills a thousand tributary canyons on its winding desert course, and hundreds of houseboats are moored at a marina near the Crossing of the Fathers, where Spanish priests forded the river in 1776. The colours of the land-

Figure 18.3 Rescue of black rhino Jane in 1993 by Mark Atkinson and colleagues from an island in the reservoir behind Kariba Dam. Jane's mother was stranded on an island during the initial drowning of the Kariba Gorge. Courtesy of Shirley and Mark Atkinson.

Figure 18.4 Glen Canyon Dam on the Colorado River in Arizona, completed in 1963. The dam holds back Reservoir Powell, 300 km long through the canyon country.

scape are arresting—the bright red rock, the improbably pale blue reservoir brushed by the wakes of powerboats. Across the lake, the Colorado Plateau rises into the haze. Downstream at Horseshoe Bend, a crowd peers over the sheer walls of an incised meander to view the regulated Colorado River, a far cry from the torrent that Powell navigated before American engineers became river gods. Glen Canyon Dam, Hoover Dam beyond the Grand Canyon downstream, and a suite of mainstem and tributary dams are a monumental testimony to engineering skill. But the Colorado is one of the world's most regulated drainages and an endangered river.

John Wesley Powell named *Glen Canyon* for its sentimental resemblance to a Scottish glen. The boats glided past tributary canyons, some with lakes dammed behind rock falls and others with winding slot canyons, barely wider than a person, where dim sunlight illuminated deep pools. Powell wrote in *Canyons of the Colorado*:

Sometimes the rocks are overhanging; in other curves, curious, narrow glens are found. Through these we climb … to where a spring bursts out from under an overhanging cliff, and where cottonwoods and willows stand, while along the curves of the brooklet oaks grow … So we have a curious *ensemble* of wonderful features—carved valleys, royal arches, glens, alcove gulches, mounds, and monuments. From which of these features shall we select a name? We decide to call it Glen Canyon.[13]

The engineering revolution ushered in an era that Powell could not have foreseen, although it was gathering momentum during his lifetime (see Chapter 20). Nowhere was big-dam fever more widespread than in the West, where 'cash-register' hydropower dams, including the Glen Canyon Dam, were often of little value for irrigation but provided funds for other projects in a basinwide accounting scheme. Drowning the canyon for purely financial reasons infuriated wilderness lovers, and naming the reservoir after a man who fought for an intelligent water policy was like rubbing salt into a wound.[14]

The dam almost came to grief at the outset.[15] After the barrier was completed in 1963, the reservoir took 17 years to fill and was newly operational when meltwater from a spring snowfall unexpectedly raised the water level. When a spillway was opened, the dam rumbled as red water laced with rock and concrete poured out, ripped away at a bend in the spillway. In the bowels of the dam, employees likened the noise to Vietnam War artillery. With springs spurting from the canyon walls, the engineers opened a second spillway and placed plywood sheets above the spillway gates to allow the highest possible level. The inflow levelled off, and a deluge that might have swept Hoover Dam into the Gulf of California was averted.

Among those who recorded Glen Canyon's final days were Katie Lee and Edward Abbey. Lee was a singer who, disillusioned with Hollywood glitz, used her nightclub career to finance her 'binge with nature'. Abbey was a park ranger, a wicked writer, and an uncompromising activist. He deplored the filthy depradations of *Slobivius americanus* and railed against park graffiti by 'obliterating from a sandstone wall the pathetic scratchings of some imbeciles who had attempted to write their names across the face of the Mesozoic'.[16]

Lee encountered log jams that beat against the rocks, dangerous quickmuds, and rainstorms that cascaded over the cliffs. She observed 'moki steps' cut in the canyon walls by Native Americans, along with petroglyphs of animal, human, and superhuman figures, and she placed her fingers in fingerprints made long before in the clay of mortared walls. Drifting through the canyon was to interrogate the river about its timeless history (realizing this, she hurled her watch into the torrent). Reluctantly she returned several times to rising 'Cesspowell': do we leave the bedside of loved ones while they are dying, she asked? Like Wallace Stegner who also visited Glen Canyon as it filled, she found that beavers had cut down every tree in the side canyons in a vain attempt to halt the rising water. Animals by the hundred starved or drowned as they scrabbled at the rock faces or, like the snakes and scorpions, crowded onto islands until the water covered them.[17] As Jared Farmer notes, the lives of the river

creatures and the river's life force were submerged in a discussion of reservoir recreation and beauty.[18]

As the reservoir filled, Abbey drifted through Glen Canyon, which he considered a portion of the original paradise. In his book *Desert Solitaire*, he wrote:

> The beavers had to go and build another goddamned dam on the Colorado. Not satisfied with the enormous silt trap and evaporation tank called Lake Mead ... they have created another even bigger, even more destructive, in Glen Canyon. This reservoir of stagnant water will not irrigate a single square foot of land or supply water for a single village; its only justification is the generation of cash through electricity for the indirect subsidy of various real estate speculators, cottongrowers and sugarbeet magnates.[19]

To Abbey, Glen Canyon was Earth in the nude, a slickrock wilderness where cottonwoods (the canyonland tree of life) rustled in the dawn breeze, a 'whispering of ghosts in an ancient, sacrosanct, condemned cathedral'. Better than most, Abbey understood the desire to reduce nature:

> Alone in the silence, I understand for a moment the dread which many feel in the presence of primeval desert, the unconscious fear which compels them to tame, alter or destroy what they cannot understand, to reduce the wild and prehuman to human dimensions. Anything rather than confront directly the ante-human, that *other world* which frightens not through danger or hostility but in something far worse—its implacable indifference.[20]

In 1975 Abbey published his crude and seditious cult novel, *The Monkey Wrench Gang*, in which four unlikely eco-heroes seek to rid the western wilderness of heavy equipment and political and economic greed. The prologue makes it clear that they are gunning for the Glen Canyon Dam, and Colorado boatman Seldom Seen Smith kneels in prayer on the nearby bridge:

> Dear old God, you know and I know what it was like here, before them bastards from Washington moved in and ruined it all ... Remember the deer on the sandbars and the blue herons in the willows and the catfish so big and tasty and how they'd bite on spoiled salami? ... There's somethin' you can

do for me, God. How about a little old *pre*-cision-type earthquake right under this dam?'[21]

Pursuit catches up with the gang before they can stuff rented houseboats with fertilizer and diesel fuel and scuttle them at the dam wall, attached to blasting wire.

Abbey's dream was to return Glen Canyon to a place 'where great waterfalls plunge over silt-filled, ancient, mysterious dams.' The land 'will breathe a metaphorical sigh of relief—like a whisper of wind—when we are all and finally gone and the place and its creations can return to their ancient procedures unobserved and undisturbed by the busy, anxious, brooding consciousness of man.'[22]

And what of Reservoir Powell? There are many downsides.[23] The trapping of silt. The loss of water by evaporation and percolation into the Navajo Sandstone of the canyon walls. Downstream into the Grand Canyon flows a programmed discharge with unnatural fluctuations of cold water, deadly for endangered river creatures. But boaters can explore unknown valleys, and sport fishing, jet skiing, and houseboating have taken over. And many obituaries in Salt Lake City newspapers honour the reservoir as a hallowed place in family memories. In Jared Farmer's view, Glen Canyon and Reservoir Powell do not cancel out, although it is hard to imagine something more beautiful under the blue water.[24]

But I find no difficulty in resisting the seductive power of the reservoir, evaporating up into the merciless desert sun. Far below the surface lies the dead, drowned bed of the Colorado River. A canyon that took six million years to form was lost in less than two decades. 'In gaining the lovely and the usable,' Wallace Stegner wrote, 'we have given up the incomparable.'[25]

Searching for the Good Dam

In his 2005 book *Deep Water*, reporter Jacques Leslie records his discussions with Thayer Scudder, anthropologist and veteran of dam resettlement projects around the world.[26] Believing that he was assisting resettlement at Kariba, he later realized that he had charted the disintegration of indigenous culture as tens of thousands of people were moved to barren land above the gorge. They had apparently always lived in the Zambezi Valley and their legends contained no stories of arriving from elsewhere. Like the Nubians along the Nile, they had depended on recession agriculture from the rich alluvial soil. By any standards, their subsequent life has been wretched.

Scudder insisted that, as the project's first and long-term beneficiaries, displaced farmers should receive a standard of living that matched their earlier life. He envisaged honest, knowledgable leaders, reliable bureaucrats, and generous donors dedicated to the 'good dam'. Really, however, he had helped to get people moved so that the dams could be sealed on time. In Scudder's view, 80 per cent of the world's big dams should not have been built, and all the others had been problematic. Kariba was one of the worst. But at the age of 71 he still felt that he might one day encounter the 'good dam'.

CHAPTER 19

BETWEEN THE DAMS: AN ELEGY FOR THE SASKATCHEWAN RIVER

North America's largest inland delta

The road to Cumberland House in Saskatchewan crosses the northern prairie, a limitless grassland little more than a century ago. Presently the road gives way to rutted gravel along the Saskatchewan River, North America's fourth longest river, which flows from its Rocky Mountain headwaters in Canada to Lake Winnipeg. A young moose trots along the road, a haze of flies around its head, and sandhill cranes drift across the big skies.

A sign heralds the Saskatchewan River Delta. Some 10,000 square kilometres in area and a habitat for millions of ducks and geese, this is the continent's largest inland delta where the big river opens out into lakes and wetlands (Fig. 19.1A). Across a bridge lies the Northern Village of Cumberland House with a hotel, a grocery store, and an ice-cream parlour, and nearby is the reserve of Cumberland House Cree Nation with its streets of compact houses.

Cumberland House was founded in 1774 as a Hudson's Bay Company trading post. The fur traders canoed upriver following routes known by indigenous peoples, who drew maps for the Europeans.[1] A century later, a sternwheeler reached Cumberland House, but the river switched its course in a major avulsion, dividing the flow into small channels that made navigation difficult.[2]

At a teacher's house rented for the summer, I meet Norm Smith from the University of Nebraska and Galina Morozova from the University of Guelph. Norm's research has placed the delta at the forefront of river science, attracting researchers from around the world. Each year Norm packs his truck with scientific gear and, peering through the windscreen cracks (a badge of distinction), makes the two-day drive from Nebraska. Norm is the first over the side of the boat with a brush axe, impervious to mosquitoes, and there is no letup in his drive to understand the river. Of an evening, he is partial to an ice cream at the parlour.

Figure 19.1A Saskatchewan River Delta, showing the main channel 250 m wide and adjoining levees and wetlands in a downflow view. Lines on the sediment bar in the foreground are accretion topography, stabilized by vegetation.

Figure 19.1B *Misty River* and Gary Carriere at a sand bar in the Saskatchewan River. Note the bar slipface at left (shadowed) and ripples on the bar top, both indicating eastward flow.

A graduate of Lomonosov Moscow State University, Galina studied with Norm and spent several field seasons on the delta, driving augers into the river plain and using radiocarbon dates to establish the river's history. Partial to jewellery and an occasional issue of *People* magazine, Galina has no reluctance about getting muddy.

At the loading dock we board Gary Carriere's boat *Misty River* (Fig. 19.1B). Gary has a lifelong experience of the river as a guide, trapper, and hunter who nevertheless can shed a tear for young animals swept away by floods. Gary knows all the channels, say the local children. He also knows subtle landmarks in the featureless mosaic of forests and rivers, where he must *feel* the land to work a trap line. Gary is a passionate advocate for the delta, and the 'Keepers of the Delta' are a dedicated group of residents who champion the river and its wildlife. A generation of kids has learned to hunt and appreciate the natural world from his lodge, and by the loading dock is a signboard where drawings by children show their feelings for the river.

Misty River swings into the channel and accelerates to 40 km an hour, as recorded by my GPS. Rivets near the bow are covered with duct tape to protect sturgeon nets from snagging, and the bow is crumpled where an unwary kid misjudged the space below the bridge at flood level. Norm, Galina and I sit facing each other across a litter of bedload samplers, a surveying tripod and level, a tarpaulin attached to long stakes, and a University of Illinois sports bag containing a spring balance, stopclocks, and sample bags. A pelican flies effortlessly ahead before wheeling off down a side channel. With its huge head and protruding bill, it seems a fantastic relic of ancient days, the last pterodactyl perhaps. We are out and away through the world.

Between the dams

The Amazon-scale Bell River drained most of Canada for millions of years before the Ice Age dismembered its drainage. About 8000 years ago as the ice sheets melted, Cumberland House lay below the waters of Glacial Lake Agassiz, where the reincarnated Saskatchewan built a new delta (Fig. 19.2).[3] The river winds through fens where blackwater channels brim with tea-coloured water. To all appearances this is a pristine landscape, barely changed since plants colonized the glaciated terrain. There is no logging, no mines, no oil production. Apart from traditional hunting and fishing, the delta is left alone.

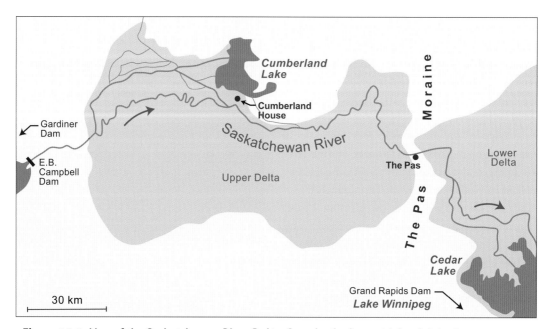

Figure 19.2 Map of the Saskatchewan River Delta, Canada, the largest inland delta in North America. The delta is divided into upper and lower areas by the glacial moraine at The Pas. Complex channel patterns upstream from Cumberland House reflect avulsions and the establishment of wetland avulsion lobes with multiple channels, with gradual consolidation of flow into a few larger channels. Cedar Lake is a remnant of Glacial Lake Agassiz, now impounded by the Grand Rapids Dam. After Smith et al., 2014.

With one exception. The E.B. Campbell Dam, completed in 1962, takes advantage of the head of water at a rapids upstream from Cumberland House, impounding a long reservoir. Turbulent water rushes through the power canal to the powerhouse turbines downstream where the flow rejoins the river. A short distance upstream is a smaller hydrodam, and some way south is the Gardiner Dam, one of the world's largest earth-filled dams, which holds back Reservoir Diefenbaker with a storage capacity greater than the river's annual flow.[4] From barrages and dams, irrigation canals siphon off water for prairie agriculture. Downstream the river cuts through a glacial moraine at The Pas to reach the lower delta, flowing into Cedar Lake, a remnant of Glacial Lake Agassiz that is now impounded behind the Grand Rapids Dam. From Lake Winnipeg, the Nelson River takes the flow to Hudson Bay through a string of dams and reservoirs.

The Saskatchewan River Delta is hemmed in by dams that divide the river into disconnected compartments. The dams have increased winter flow for power generation but have suppressed the spring and summer floods that once recharged the wetlands. Beaver are moving into the river because the wetland lakes now freeze too deeply for them to survive. The reservoirs trap the nutrient-rich sediment that once fertilized the delta, and fluctuating releases from the dams may change river levels by as much as two metres in a day, drowning the nests of piping plover.[5] Over the past century the river's discharge has dropped by about a quarter due largely to irrigation takeout and evaporation in this semi-arid region. Climate change, too, affects the river's water budget.[6]

In the mid-1950s, the federal Prairie Farm Rehabilitation Administration (PFRA) commissioned a study to explore draining the delta for farmland. Working in the dead of a prairie winter, Dutch surveyors mapped the delta: better than anyone else on Earth, they understood how subtle gradients influence water flow. Surveying along the frozen river, they obtained precise elevations and hammered nails into the trees for benchmarks. They axed holes through the ice, lowering weighted lines to measure the depth, and they collected riverbed samples. From their measurements they plotted channel and wetland cross-sections on rolls of paper 20 m long.

A decade later, the dams were built and only a small corner of the delta was drained for agriculture. The cross-sections lay unregarded in a Winnipeg office

for nearly half a century, and they constitute a unique pre-dam record of the delta channels, allowing us to compare their dimensions with the modern channels.

Hungry water at the power canal

Misty River heads upstream to the power canal of the E.B. Campbell Dam. Ducks rocket up ahead while a canny heron flaps lazily over the bank and is gone in an instant. A merganser stands courageously out in the river as her nine chicks scuttle into the shallows. Black dragonflies are hawking over the water and the sandy banks are stippled with the holes of nesting swallows.

Norm is scanning the banks, comparing them with an aerial photo that shows the position of PFRA's channel cross-section C102. There is a benchmark somewhere nearby, but the banks and looming forests all look alike until Norm points to a group of prominent firs on the photo, still visible above the poplars. Finding a benchmark is to search for a single tree in the forest, but this benchmark has coordinates programmed into my GPS. As I leap ashore, the GPS freezes under the canopy of trees, cut off from its satellite signal. Meanwhile Gary has walked straight to an old large tree and found the benchmark. 'A triumph for Indian science', says Gary.

Moose footprints, enormous irregular depressions, cross a muddy slough on the bank. Norm sets up the tripod and levelling telescope nearby, and I hold the surveying rod. The mosquitoes rise joyfully in grey clouds, whining and biting. It is not possible to hold the rod still and slap them away, and for a moment I envy the Dutch surveyors working at forty below without insects. After what seems an eternity, Norm has his elevation measurement on the benchmark for an exact comparison with cross-section C102.

How have the channel width and depth changed since the PFRA survey? We set up the tarpaulin vertically on the bank where it can reflect signals from a laser rangefinder on the boat: the trees won't give a coherent reflection. Norm is hacking away low shrubs for a line of sight. We pile into the boat and Gary drives towards the far bank, 300 m distant. Every few metres, he expertly holds the boat steady in the current, and I focus the rangefinder on the tarpaulin while Galina holds an acoustic depth-meter in the water. We sing out the readings for Norm to record—loudly because he is a little deaf.

Gary guns back across the river and we take down the tarpaulin, repacking the boat (Fig. 19.3A). It has taken 40 minutes with very simple equipment to make our cross-section. A dead elk drifts by, its body and antlers damaged by wolves or a bear. A pelican skulks near a bank with its wing hanging, not likely to survive long.

Misty River bucks through turbulent boils of water below the powerhouse, where hundreds of pelicans, gulls, and cormorants are snapping up the fish that emerge stunned or dead from the turbines. Norm unpacks something that looks like a hollow torpedo but is actually a steel gatepost discarded at a building site (Fig. 19.3B). Thrown over the side on a rope, the torpedo drags along the bottom, capturing boulders covered with weed. After repeated trawls, it is evident that the reach is covered with boulders for a long way below the power canal. In 1955 the Dutch surveyors found only sand, long since removed by the turbulent flow.

Lunch is eaten in the boat to avoid the insects. Gary has a sandwich of 'organic moose meat' and a flask of coffee. A moose is swimming the channel, panicked by our sudden appearance, and we are so close that I could leap onto its back. Gaining the bank, the big animal crashes off through the forest.

We head back to Cumberland House, skirting bars of rippled sand. Galina sits in the bow, drinking in the air as it rushes past. 'I feel completely alive out here', she says. Thunderheads edged with light rise above the dark forest, and the rain is suddenly upon us out of a still sky. Gary puts on rubber coveralls, Norm hunkers down under a tarpaulin, and Galina sports a pink raincoat. I am captivated by the shimmering water as the droplets strike.

The rain clears at last. At every bend in the river a bald eagle stands sentinel on a dead tree, some guarding nests where young birds are rustling about. An eagle feather twirls on the river surface. A young bear peers from the bushes, rushing away with a crackling of branches, and a wolf drinking at the river is caught unawares. An armada of small boats is heading to the delta to fish for walleye along channels fringed by tufted grass and the wild rice *Zizania*. We knock off rice-rich heads into the boat as we pass using the traditional 'canoe-and-flail' method of harvesting with paddles or poles. The river broods.

Figure 19.3A Surveying with *Misty River* on the Saskatchewan River.

Figure 19.3B Norm Smith with weed-covered boulder from a sample tube below the E.B. Campbell Dam, where 'hungry water' emerging from the powerhouse turbines has stripped the riverbed of sand.

After dinner Norm works with the cross-section for line C102. He passes a page of numbers to Galina who enters them into a computer spreadsheet along with our measurements for the day. Presently she calls us over to compare the two cross-sections. The modern channel is much deeper and wider than it was in 1955.

At the river's end

The field season nears its finish, and *Misty River* takes us downstream to the town of The Pas, passing a slowly circling whirlpool tens of metres across. En route there is an unsuccessful search for a benchmark described as '6-inch spike 1.0' above ground in 8" diameter-blazed maple 30' south of winter road and on the north bank of Saskatchewan River approximately ¼ mile east of island in most northerly bend of river'. We find the winter road and a cabin with 'Please use other door' written on the only door. But the maple and its spike have long since toppled into the river.

Norm's orange ballcap with its logo 'Nebraska Hunter Education Programs' blows off into the river and has to be retrieved. A metre-long sturgeon leaps clean out of the water by the boat and slaps down on its side, an ancient fish now in trouble across Canada, its population broken into small segments between dams.[7]

Norm samples the river sediment from the bridge at The Pas, lowering a temperamental sampler christened 'Beelzebub' and drawing it back over the railings filled with sand. Is he trying to catch Jaws? a passerby asks, and the police come by for a friendly chat, alerted that a man with a rope is attempting to do away with himself on the bridge. A phone message from Gary comes in that I am lost, but not in the forest: he is at the grocery store while I am at a coffee shop.

We need sediment samples down to Cedar Lake, but not even Gary has visited this remote place. As we set off early in the morning, a flock of black terns keeps pace like escorting outriders and a Canada goose overtakes the boat with steady wingbeats, a match for the powerful engine. Beavers slip into the water from their lodges and a group of otters, one carrying a fish, glide past.

For sediment sampling we use another of Norm's inventions—a Mexican espresso can with the top removed, weighted with railroad spikes and wrapped around with duct tape. Hurled over the side on a rope, the can is pulled along the bottom and fills with sand, wood fragments, and clams. Lumps of grey mud mark the ancient clay of Glacial Lake Agassiz. Sometimes the can is empty, the riverbed swept clean of sand.

This is my first visit to the end of a major river, and there is a magical quality to this place and time. A few pelicans and terns drift by, and red-winged blackbirds chuckle among the grasses. Loons pierce the stillness with their primitive calls. A fishing cabin is strewn with flood debris and the owner has left the door open (as I like to think) to allow the swallows to build their mud nests under the roof. The grassy banks have no mosquitoes: even they have abandoned this remote place.

Beams of sunlight play through the clouds onto Cedar Lake, seemingly drawing up moisture to the sun

as English villagers once thought. At the farthest limit of the river stands a lone white tree with an eagle's nest, where a solitary chick watches us stolidly while the parent circles overhead. The only sound is the rustling of the grass, and my hearing seems strongly intensified. A clock ticks faintly: a timer has chosen this moment to spring to life inside the University of Illinois sports bag.

The last sample is taken and the boat drives westwards to The Pas up a brilliant ray of setting light. Deadhead logs, skulking below the surface, run under the boat, and we are delayed by an airlock in the motor that takes a while to fix. But it is one of those serene evenings when it seems that nothing can go wrong, as Robert Browning expressed it in his poem *Pippa Passes*: 'God's in his heaven—All's right with the world!'

Over the next months Norm pulls the measurements together. It is clear that the E.B. Campbell Dam has radically affected the Saskatchewan River.[8] With its sediment load trapped in the reservoir, an energetic flow starved of sediment roars out from the powerhouse, cutting back the banks and eroding sand from the channel floor to leave a residual armour of boulders. The 'hungry' flows have enlarged the channel for 80 km below the dam, excavating some 35 million cubic metres of sediment—a volume equivalent to decades of rock removal in major mining areas. With more space in the channel for floodwater, less water spills over to replenish the wetlands, already reduced by smaller flood releases and the loss of water to irrigation and evaporation.

To a visitor, the delta seems a marvellous representation of unspoilt nature. But no river can absorb such a change, and Gary knows more than anyone how profoundly the delta has been affected. Even Cedar Lake at the river's end is a reservoir. In 2009 the World Wildlife Fund identified the South Saskatchewan as Canada's most threatened river, calling for flow patterns to support the natural ecosystem.[9] And the flood levels: are they determined by Acts of God or Acts of Dam Control Engineers?

Downstream at Grand Rapids (Fig. 19.4), the cascading water had been audible from a great distance, and infants taken by canoe through the rapids were believed to gain special strength of body and spirit. When the dam was completed in 1968, the rapids were suddenly 'turned off'. In the uncanny silence, excited children looked for old guns on the dry riverbed and

Figure 19.4 Grand Rapids Dam at the downstream end of Cedar Lake, with Galina Morozova on the earth dam and Lake Winnipeg beyond.

caught fish thrashing in stagnant pools.[10] The river song had been eliminated, and the fishery collapsed when fish could no longer pass the dam. No alternative employment was provided. Before the dam was built, the community's children couldn't remember being hungry. Hunger was widespread afterwards.[11]

A terrible and vast reduction of our entire world

From Lake Winnipeg, the Nelson River runs out to Hudson Bay, as does the Churchill further north (Fig. 9.2). An enormous hydropower scheme governs these rivers, and a dammed lake in the Churchill system flooded thousands of kilometres of river shoreline and reversed the flow into the Nelson system to boost power generation.[12]

The dams were not responsible for every problem, but the land became *terra incognita*, an unrecognizable landscape. Deaths resulted when boats collided with floating debris, fish were contaminated with mercury, and large game animals moved away. An intimately known homeland became unpredictable and dangerous. Out of sight, northern rivers and their ecosystems and communities paid the penalty for the human enterprise. As community member Ramona Neckoway commented, the only creatures allowed to build dams should be beavers.

In the James Bay lowlands to the south with their immense hydropower projects, records from the time of construction underscore the contrasted world views at play.[13] For government leaders, hunting and trapping were an occupation to be expressed financially. For the Cree, they were part of an intimate relationship with an ever-changing landscape that provided life, traditions, and spiritual resources. The unwillingness of community members to answer questions precisely was a refusal to think about the land in numerical terms. 'When you talk about money' said one hunter, 'it means nothing, there will never be enough to pay for the damage that has been done. I'd rather think about the land and think about the children'.[14] Matthew Coon Come, later the National Chief of the Assembly of First Nations in Canada, was instrumental in halting a hydro project, telling potential users, 'This is what I want you to understand: it is not a dam. It is a terrible and vast reduction of our entire world. It is the assignment of vast territories to a permanent and final flood'.[15]

Sigurd Olson canoed down the upper Churchill shortly before big dams sealed the northern rivers.[16] An academic Dean, Olson gave up the university for wilderness guiding and writing, negotiating whirlpools and lakes whipped to a frenzy by storm-force winds. Olson propped up in the canoe a quotation from an old monk in *The Brothers Karamazov*, and he meditated on it as he paddled: 'Love all God's creation, both the whole of it and every grain of sand. Love every leaf, every ray of God's light. Love animals, love plants, love each thing. If you love each thing you will perceive the mystery of God in things'.[17] The Churchill was a great free river, but he doubted whether even this vast wilderness could survive.

Back home, I am in the mall to buy a smart phone. The sales consultant displays a range of devices and explains the financial packages available. My eye drifts vaguely to a large screen where images of a wetland are scrolling through. Suddenly, the face of Norm Smith fills the screen, wearing the same orange ball cap that I fished from the Saskatchewan River a month before. Norm is followed by Gary Carriere, *Misty River*, a pelican, and the E.B. Campbell Dam.

I leave the store with an iPhone, having signed on for something—I hardly know what. I am lost in the complex waterways of the twenty-first century. In my hand is an electronic appliance run by hydropower. In my heart are the channels of the Saskatchewan River.

CHAPTER 20

WITHOUT SPOILING THE LAND: RIVERS AND AGRICULTURE

Living soils along the Amazon

In 1870, Cornell geology professor Charles Hartt took his students to Brazil.[1] They travelled up the Amazon between walls of foliage to reach rock outcrops with views across the 'inland river sea' where, at times, the alligators swarmed like tadpoles. The party fought through sword grass that shredded their clothes and swarms of biting flies that blackened and infected the skin. 'These pests disappear at night, and, for a change, mosquitoes take their place,'[2] Hartt commented drily. But Hartt loved the Amazon landscape with its unknown geological history, finding Coal-Age plant fossils for John William Dawson back in Canada.

Hartt found 'all corners of time' for the indigenous people. They had occupied caves some 12,000 years earlier, foraging for fruits and river creatures, using stone tools, and printing their hands in rock paintings.[3] Along the river near Santarém, he watched village women moulding clay vessels, quaintly noting the similarity between the patterns that they created and the embroidery of wealthy ladies back home.

Hartt was especially intrigued by the black soil on the bluffs above the Amazon, which he attributed to long-vanished human settlements.[4] One of Hartt's students, Herbert Smith, dug out pottery from the soils at Santarém, along with clay figurines of vultures, frogs, caymans, and monkeys. The bluff soils, he thought, owed their richness to centuries of treatment with kitchen refuse. He visited farms run by *Confederados*, Americans who had emigrated after the Civil War and had selected sites with dark soils after a lengthy river survey.

Termed *anthropogenic dark earths*, these soils are now known to cover some 18,000 square kilometres of Brazil and Guyana, an area the size of a small European country.[5] Interest in the dark earths was revived by Dutch soil scientist Wim Sombroek, whose family had

Figure 20.1 Anthropogenic dark-earth soil on the Amazon River bank at Claudio Cutião, Brazil, occupied from 1,550 to 550 calendar years Before Present. The projecting fragments are pottery shards. Bands on the staff are 10 cm long. Courtesy of Myrtle Shock.

survived starvation during the Second World War by subsisting on produce from a garden enriched over generations with fireplace ash.

The dark soils are commonly about half a metre thick and date back at least 6000 years. They contain three times more carbon than the normal forest soils and are rich in charcoal (Fig. 20.1). The indigenous people in settled communities had worked lovingly with their soils, and surviving settlements still mix burnt plant ash with household cooking waste, bones, and manure. Aided by communities of microbes, worms, and ants, the dark soils maintain their fertility almost indefinitely and regenerate within months of cultivation. In places, flat areas where houses or yards once existed are surrounded by ring-shaped mounds of anthropogenic soil—a 'middenscape' formed from kitchen waste.

The Amazon cultivators had lived sustainably with river soils for thousands of years, something that has eluded almost all human societies. As conservationist Aldo Leopold noted in a 1938 essay,

> the standard paradox of the twentieth century: our tools are better than we are, and grow better faster than we do. They suffice to crack the atom, to command the tides. But they do not suffice for the oldest task in human history: to live on a piece of land without spoiling it.[6]

'Rain follows the plough': agriculture west of the hundredth meridian

John Wesley Powell became a leading scientific figure in Washington after his Colorado River journey. As Director of the recently founded Geological Survey, he moved to implement a radical water policy for agriculture in the West. Having grown up on farms, Powell knew what it took to work the land. He also knew that less than 20 inches (~500 mm) of rain fell each year west of the 100th meridian, considered the threshold for agriculture without irrigation.

Powell's 1878 report on the western rivers and drylands stated that so little water was available that homesteaders could not survive with a standard eastern allotment, surveyed on a grid system.[7] Each failed homestead would represent a loss of livelihood and perhaps life, and cattle barons and railroad companies would reap the rewards. Instead, each family should receive a plot of irrigable land and a much larger plot for grazing, and irrigation systems should be based around the rivers and be run cooperatively, like those of the Native American and Mormon settlements with which Powell was familiar. He proposed withdrawing land grants pending surveys of irrigable river land where farmers could be assured of success.

Powell knew that rainfall was erratic, but others thought differently. Midwest settlement in the 1870s had taken place against a backdrop of good rainfall, and a peculiar folklore held that 'rain follows the plough'.[8] The supposed increase in rainfall was variously ascribed to the absorptive power of cultivated soil, tree planting for windbreaks, telegraph lines and railways, or just to mysterious natural forces. The plough was an 'unerring prophet', proponents declared, and the West could be settled without arduous dryland farming. Bountiful prospects were the manifest destiny of the West. Deserts persisted through human lethargy.

Encouraged by this view, vested interests across North America pressed for uncontrolled settlement on survey-grid farmsteads. Powell's carefully planned irrigation programme was gutted by a powerful Washington elite, and he was heckled when he declared at a congress, 'I tell you gentlemen, you are piling up a heritage of conflict and litigation over water rights for there is not sufficient water to supply the land'.[9] He resigned from the Geological Survey and died in 1902. In the words of Wallace Stegner,

> It was the West itself that beat him, the land and cattle and water barons, the plain homesteaders, the locally patriotic, the ambitious, the venal, the acquisitive, the myth-bound West which insisted on running into the future like a streetcar on a gravel road.[10]

As Powell predicted, droughts across the West led to high rates of farm failure even before the Dust Bowl.[11] On the Columbia Basalt Plateau in the 1920s, J Harlen Bretz found abandoned farms and schools with books and maps scattered about.[12] It is unclear whether Powell, for all his sagacity, foresaw the eagerness for westward expansion and the sheer resilience of the farmers. At root was a culture that viewed land and rivers as resources to be exploited, without reverence and restraint.[13] But in Stegner's view, Powell would have opposed a political vision for rivers based on engineering feasibility. As Marc Reisner comments,

> [BuRec] set out to help the small farmers of the West but ended up making a lot of rich farmers even wealthier at the small farmers' expense … We set out to tame the rivers and ended up killing them.[14]

Aldo Leopold jumps out over river management

In 1922, conservationist Aldo Leopold canoed to the Pacific through the Colorado delta in Mexico.[15] More than a decade earlier, he had signed on as a Forest Assistant in Arizona with his own horse and regulation pistols, narrowly surviving a lightning bolt that drove a huge wood splinter into the ground at his feet, where it hummed like a tuning fork. Later, as Professor of Game Management at the University of Wisconsin, Leopold's resource training began to conflict with his feeling for the natural world, and 'wildlife' began to replace 'game' in his writing. Like Powell, he attempted to chart a sustainable path for agriculture.

Through the nineteenth century across the northern prairies, ploughs broke through the tangled grassland roots with a sound like distant screaming.[16] Steam-powered reapers and threshing machines were welcomed by Victorian countrymen like Richard Jefferies, who knew the drudgery of farm labour. But soon more efficient gasoline-powered tractors and combines rapidly expanded the ploughland, and wheat and new crop varieties replaced the drought-resistant prairie grasses across large areas.

Wisconsin, like much of western North America, was soon experiencing the Dust Bowl, which devastated the plains for a decade from the early 1930s.[17] Crop failures left the fields stripped of vegetation, which amplified the frequent droughts, and the defenceless prairie soils were whipped up into unprecedented dust storms. On Black Sunday in 1935, a vast cloud of dust swept eastward, raining down on cities and ships far out to sea. Rabbits and small birds died in their thousands, and fish succumbed in the rivers. People were marooned in an opaque darkness, many donning facemasks so that city streets resembled a gas attack on the Western Front. Some died of suffocation and many from respiratory ailments. According to a reporter, Lady Godiva could have ridden through the streets without even the horse seeing her.

At the height of the Dust Bowl, some 850 million tons of soil were blown away each year, approaching the Amazon's annual sediment load. Some rivers with mountain headwaters maintained their flow, but others were reduced to seepage around dam walls and effluent from sewage plants. Navigation was suspended on the upper Mississippi. Powell's ideas were resurrected and debated as dispossessed farmers moved west to California.

Alarmed at the pace of landscape change, Leopold became an influential voice. The close working relationship between rivers, soils, plants, animals, and humans was fragile: it could be disrupted in a moment. He engaged with farmers about building rivers and wildlife into a sustainable system for prairie agriculture.

But in the halcyon days of 1922 the Colorado delta was still untouched. Leopold and his brother paddled through a maze of vegetated creeks where the delta creatures seemed in a constantly festive mood and flocks of sandhill cranes circled overhead:

the river was nowhere and everywhere, for he could not decide which of a hundred green lagoons offered the most pleasant and least speedy path to the Gulf. So he travelled them all, and so did we. He divided and rejoined, he twisted and turned, he meandered in awesome jungles, he all but ran in circles, he dallied with lovely groves, he got lost and was glad of it, and so were we. For the last word in procrastination, go travel with a river reluctant to lose his freedom in the sea.[18]

Leopold wisely never returned to the delta, aware that the green lagoons were now raising canteloupes. As BuRec Commissioner Floyd Dominy stated, 'The unregulated Colorado was a son of a bitch. It wasn't any good. It was either in flood or in trickle.'[19] While Reservoir Mead rose behind the newly closed Hoover Dam, no freshwater reached the delta for six years, a scenario repeated over the 17 years that the Glen Canyon Dam took to fill Reservoir Powell. Today, dams and diversions for irrigation, industry, and western cities have taken three-quarters of the water that once dallied through the delta. Considered a dead wasteland of mudflats by the 1970s, the Colorado delta has been making a modest comeback.[20]

In his later years, Leopold focused especially on the plight of rivers, contesting plans to dam the Wisconsin River. He deplored the short-sighted engineering that fixed the channels rather than addressing the 'deranged watersheds' generated by farming, which weakened the soil and silted up the reservoirs.[21] 'The building of a power dam', wrote Leopold, 'is an act of violence on nature and it is up to somebody to prove a dam will make the river more valuable than it is without it.'[22] 'My own impression,' he commented, 'is that river "development" is now going to such lengths as to leave people like myself virtually at odds with society. I can't ride in a car going in the opposite direction from my destination. I'd rather jump out.'[23]

In 1934 Leopold purchased a worn-out farm by the Wisconsin River, sprucing up a chicken coop to make a cabin. The family planted tens of thousands of trees, practising his dictum that 'conservation is something a nation learns'.[24] *Aldo Leopold's Shack* still stands as a testimony to his vision, with a hand pump by the

Figure 20.2 Aldo Leopold's Shack by the Wisconsin River, purchased in 1934.

front door and a rough stone fireplace with cast-iron pots (Fig. 20.2). It was here that Leopold wrote *A Sand County Almanac*, named for the sands of Glacial Lake Wisconsin, which set out his conservation blueprint for a 'land ethic'. Seeing the migrating flocks of sandhill cranes over the shack, he was reminded of their trumpet-like call on the Colorado Delta:

> We and they had found a common home in the remote fastnesses of space and time; we were both

back in the Pleistocene. Had we been able to, we would have bugled back their greeting. Now, from the far reaches of the years, I see them wheeling still.[25]

A new river on the Pampas in Argentina

It had rained heavily that night, and the clamour of the water continued with daylight.[26] Stepping outside, Argentinian farmer Ana Risatti was shocked to find that the roar of water was coming from a deep gully that had not been there the day before. The land just beyond her fenceline was a miniature Grand Canyon filled with turbid water, earth, and trees.

Little more than a decade later, Ana's gully has become Río Nuevo (New River), 25 m deep and 25 km long (Fig. 20.3). The river knifes through the former Pampas grasslands, threatening towns and roads, and it has covered the fields with flood sediment. The trees that formerly stabilized the soil and used much of the rain have made way for seasonal crops of soya beans and maize. The water table has risen, and the soils have collapsed and become saline. With the new crops unable to protect the land, the total length of streams

Figure 20.3 'Río Nuevo' cutting across agricultural land near Villa Mercedes in Argentina. The newly formed channel is about 100 m wide, with steep scalloped walls cut into the floodplain and pale areas of salt efflorescence on the sediment bars. Maps data: Google, CNES Airbus.

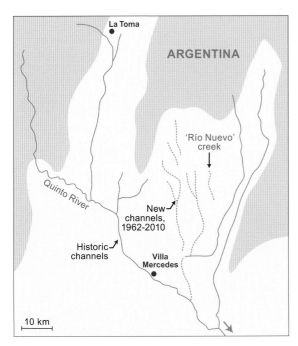

Figure 20.4 Recently formed channels in the 'Río Nuevo' area of Argentina. The length of channels in this area has tripled in half a century. After Contreras et al., 2012.

on the river plain has tripled over the past half century (Fig. 20.4). Reclamation will require perennial and salt-tolerant crops with deeper roots, but large areas will need to be reforested. By then, however, the large agribusinesses will have moved on.

Manufactured river landscapes

It is 2013 and I am flying over the midwestern plains of America. Out to the horizon are patterned fields with circles where pivoting sprinklers are pumping up groundwater for irrigation. Spurred by biofuel in-

centives, midwestern farmers have converted most of the remaining grasslands and wetlands to fields at rates not seen since before the Dust Bowl and comparable to rates of rainforest loss in South America.[27]

Below me a meandering river comes into view. The curved furrows of the ploughed fields run right into the meander bends and up to the banks, restricting the river to a thin line. No shrubs or grassland adjoin the river, no uncultivated margin. The river is caged as surely as a zoo animal, disconnected from the adjoining plains in a manufactured landscape.[28]

Worldwide, some 70 per cent of global water use is for irrigated agriculture.[29] The Amu Darya and Syr Darya feed the Aral Sea, once the world's fourth largest lake by area, but largely dried up due to irrigation takeout. In West Africa, dams and barrages for irrigation along the Niger, Senegal, and the rivers that feed Lake Chad have dried up the lakes and wetlands. Millions of displaced people have moved to cities and, with no livelihood, have joined insurgent groups such as Boko Haram or have ventured the dangerous migrant routes to Europe. Many of the region's problems stem from the mismanagement of water in a drought-prone region.

America's second longest river, the Rio Grande, runs through the famous *bosque*, a rich riparian ribbon like the jungle of the Jordan. But the Rio Grande is seriously at risk from irrigation takeout, and the river fails to reach the sea in some years. Aldo Leopold wrote an inscription for his son on the flyleaf of his book *Game Management*, remembering their time on the Rio Grande:

> The greatest fortune I can wish you is that you and your son may someday find such a river, and that there may still be mallards to fly when the dawn wind rustles in its cottonwoods.[30]

CHAPTER 21

LONDON'S BURIED RIVERS

John Snow removes the pump handle

Leaving the elegance of London's Mayfair, the visitor plunges into Soho with its maze of narrow streets. A tourist shouts, 'Hey, man, it's Carnaby Street!', but I follow the road where Karl Marx lived in the 1850s to reach Broadwick Street, formerly called Broad Street.

And here is the John Snow pub (Fig. 21.1) where partygoers have spilled out onto the street on a summer evening, a happy mixture of the well dressed, the partly dressed, and the couldn't-care-less dressed. Beneath an inn sign of the famous doctor, who was a teetotaler, passionate encounters are in progress, selfies are legion, and I am by some decades the oldest person present.

But where is the famous Broad Street pump? A sign directs me to a granite kerbstone where the pump once stood, 'associated with Dr. John Snow's discovery in 1854 that cholera is conveyed by water'. The rock is camouflaged by beer stains. As I stoop down, a flipped cigarette butt bounces off my shoulder.

On 31 August 1854 there occurred what Snow described as Britain's most terrible outbreak of cholera.[1] Seven hundred people died, many within hours of contracting the disease, and some were nursed in a nearby hospital by Florence Nightingale. The population fled in terror, and the alleyways were swiftly deserted. A meticulous man, Snow plotted the locations of the deaths, finding that 500 people had died within a short distance of the Broad Street pump. He chased down the exceptions. People from further away had liked the sparkling Soho well water, and children had walked from some distance to a school near the pump. In contrast, people at the nearby brewery and workhouse had survived, drinking beer or taking water from a deep well.

The parish Board of Guardians met in a gloomy session with no notion of what to do. Suddenly a stranger was announced. In a modest speech, John Snow laid out his findings and advised the Board to remove the pump

Figure 21.1 John Snow pub in Soho, London. The pump that provided contaminated water, linked to the 1854 cholera epidemic, was located below the pub sign with Dr Snow.

handle immediately. Incredulous at first, they followed his advice and the epidemic was over, although Snow noted that fatalities declined after the residents fled.

The Broad Street well was close to a shallow cesspit that had leaked faecal bacteria into the drinking water. From long-term medical statistics, Snow discovered that customers drawing unfiltered city water from the Thames suffered 114 deaths per 100,000 served, whereas customers drinking water from an unpolluted reach upstream were unaffected. Eels found in house water pipes indicated how carefully a company filtered its river water. Snow was vindicated in 1884 when cholera was linked to a water-borne bacterium.

On the pub wall is a copy of Snow's map with black squares clustered near the pump. I drink a glass of tap water in honour of his enduring contribution to the populace and rivers of London.

The decline of the Thames

The Thames has followed its present course for nearly half a million years after advancing ice pushed the river south.[2] Entrenched in the impervious London Clay, the river's tributaries may become raging torrents after rainstorms, forced to the surface from the underground passageways where, over the years, they have been buried alive (Fig. 21.2).

Roman legions forded the river at Westminster, where the Thames was shallow. The invaders built a fortified town where the modern City of London stands (Fig. 21.3), supplied by the Walbrook which was soon choked with trash that included Roman coins. For the next thousand years, London was a city of flowing streams and springs of sweet water, with water mills where the tributaries dropped steeply to the Thames. In Chaucer's *The Canterbury Tales*, written in the late fourteenth century, pilgrims en route to Canterbury stopped at a watering place at Earl's Sluice, where the noble knight drew the short straw and began the bawdy tales. As the city grew, butchers washed out the entrails of animals in the rivers, and tanners and skinners set up trade along the banks. The stench from Stamford Brook interrupted the religious duties of Carmelite friars, and local wags rechristened Sherbourne Lane as Shiteburn Lane.

The River Fleet declined to a polluted ditch and then to an underground drain. In Ben Jonson's vulgar satire *On the Famous Voyage*, written in *c.*1612, two men row up the Fleet in an open boat, braving the stench of sewage and alarming sounds from latrines overhead. Diarist Samuel Pepys waded through excrement in his basement, which had received the overflow from his neighbour's cesspool,[3] and an escaped pig lived so well in the sewage system that its value quadrupled in five months. After the Great Fire of London in 1666, Christopher Wren built the elegant Fleet Canal in an attempt at beautification, but it rapidly filled with rubbish. In 1678, writer and gardener John Evelyn described the Thames as the 'Sweetest River in the World'—either a reference to the river upstream or a burst of deluded nostalgia.

There was only one thing to do: cover the tributaries and seal in the stench. Confined to brick culverts, the rivers drifted out of knowledge, although criminals often dropped their victims through trapdoors into the channels below.

By the early nineteenth century, the growing metropolis boasted some 200,000 privies above cesspools where the liquid waste filtered into the groundwater.[4] Sewer gases exploded, once destroying three houses above the buried Fleet. To relieve stress on polluted groundwater, house drains were permitted to empty into sewer pipes that fed into the rivers, and the indoor water closet with its flushing and refilling system contributed to the problem. The many small cesspools became a single large cesspool—the River Thames. Miners constructing the Thames Tunnel contended with foul gases leaking from the sewer-like river above. Alexander von Humboldt descended in a diving bell to explore the opaque river bed, coughing up blood (Isambard Kingdom Brunel was unaffected: 'It is a Prussian privilege', said Humboldt).[5]

Disease spread through the polluted city. Charles Dickens, editor of the weekly magazine *Household Words*, ran articles on sanitation after the 1854 cholera outbreak. The magazine considered that typhus was a deadlier enemy than the Czar and discussed inept government solutions, commenting sourly:

> It comes home to the minister of state. He may sacrifice sanitary legislation to the first comer who attempts to sneer it down, and journey home to find the grateful plague sitting in his own hall ready with the only thanks that it can offer.[6]

In 1855, physicist Michael Faraday noted that the Thames was an opaque brown fluid with an appalling smell. Pieces of card dropped into the water became invisible an inch below the surface. He concluded in a letter to the *Times*,

> If there be sufficient authority to remove a putrescent pond from the neighbourhood of a few simple dwellings, surely the river which flows so many miles through London, ought not to be allowed to become a fermenting sewer.[7]

Matters came to a head during *The Great Stink* of 1858.[8] So hideous was the river's smell that, in the Houses of Parliament, curtains impregnated with chloride of lime were hung over the windows. Chancellor of the Exchequer Benjamin Disraeli rushed from the Chamber bent over with a handkerchief pressed to his nose. Following him was William Gladstone with his face thrust into a cloth and a baronet coughing up phlegm. The discomfiture of the nation's representatives

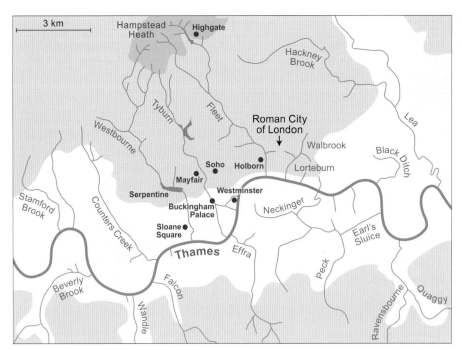

Figure 21.2 Map of the River Thames, London, with courses of tributary rivers, now mainly underground. After Barton, 1962.

Figure 21.3 River Thames in London, 250 m wide at this point, with the dome of St. Paul's Cathedral in the area of the original Roman town. The Fleet River enters the Thames through a grille below Blackfriars Bridge at centre left.

accomplished what no amount of suffering among the lower classes had achieved, and Parliament allotted three million pounds to the Metropolitan Board of Works to clean up what Disraeli called a pool of unbearable horrors.

How was life in London before the lost rivers were cleaned up? I have on my desk a leather-bound volume of *The Illustrated London News* for 1855. The Crimean War with Russia is at its height with an investigation into the ill-fated Charge of the Light Brigade, and the paper illustrates Florence Nightingale riding to the front. My hometown of Tiverton gives 'three decided groans' for the Czar. On the home front, leaden combs will darken your hair and Newfoundland cod-liver oil will improve your health ('Newfoundland affords no other livers for making it but those of the true cod'). French chemists have discovered a metal called aluminium, although its economic value is uncertain.

But the column of Metropolitan News illustrates London's crisis of poverty and health more dramatically than any Charles Dickens novel. In an unusually severe winter, the Thames has frozen over, ships are trapped in the ice, and 1200 dock workers facing starvation are rioting for bread. More deaths than births are reported week after week, many from diseases linked to sanitation, and more than half the dead are under the age of 20. With sewers filled with garbage and little water for cattle, the old Smithfield Market by the Fleet has closed. One hundred million gallons of water are extracted from the polluted Thames daily, for human and industrial use.

How an engineer saved London

The hero of the hour was Joseph Bazalgette, Chief Engineer of the Metropolitan Board of Works, who is credited with saving more lives than any Victorian physician.[9] From his office on Soho Square, Bazalgette drew up plans for east–west sewers below the city, intercepting the underground rivers that ran north–south. Pumping stations brought the sewage to the Thames downstream of London, from where the tide took it out to sea. The lost rivers became storm drains that allowed overflows to reach the Thames.

Bazalgette's plan required the construction of more than 130 km of sewers under the city. To accommodate the low-level sewer, he reclaimed parts of the river as embankments, narrowing the flow but making space for underground railways. Critics thought that a sewage system discharging downriver was vulnerable to attack by hostile fleets, but others felt that London's sewage discharge would deter the most resolute enemy. Bazalgette's system was officially opened in 1865 and completed a decade later. After 1866 no epidemics of cholera or typhoid were reported.

Out of sight downstream, pumps discharged millions of gallons of sewage a day into the river. In 1878 the paddle steamer *Princess Alice* with 900 passengers collided with a larger vessel and sank within minutes, shortly after the day's sewage release.[10] Few passengers made it to land, and most of the survivors died within days from bacterial diseases. In 1887 the city accepted Bazalgette's proposal for sludge boats to take the solid waste out to sea, a procedure that continued for more than a century.

Following the lost rivers

London's rivers did not cease to exist because they were forced underground. Their former courses can be discerned from the lie of the land and from placenames linked to wells and water, as well as from historical locations in novels such as those of Dickens. Many reported hauntings in London come from houses near the former waterways, perhaps the ghostly sounds of the lost rivers flowing underground.[11]

The Fleet rises on Hampstead Heath and flows out through Hampstead Ponds past the graves of Michael Faraday and Karl Marx at Highgate Cemetery. In Dickens' *The Pickwick Papers*, Samuel Pickwick communicated a paper entitled 'Speculations on the Source of the Hampstead Ponds, with Some Observations on the Theory of Tittlebats' (stickleback fish). He felt pride in presenting his Tittlebatian Theory; it might be celebrated, or it might not (a cry of 'It is', and great cheering). But Pickwick's river explorations were diverted into a backwater when, accused of breach of marriage, he was incarcerated in the Fleet Prison for debtors. The river can be heard through the drain grids near Smithfield Market, running underground below Holborn Viaduct (Fig. 21.4A) where wagons once struggled up from the ravine. A black cascade from the Fleet flooded the underground railway during construction, and the aged Prime Minister Lord Palmerston excused himself from the opening, wishing to

Figure 21.4A Line of the Fleet River running underground below Holborn Viaduct, which spans the former ravine.

Figure 21.4B Tyburn Brook running through the basement of an antique shop in Mayfair.

stay above ground for as long as possible.[12] Dickens placed the dwelling of Fagin, who apprenticed Oliver Twist in street crime, in the slums of Saffron Hill on the former riverbank, and Old Seacoal Lane and New-castle Close date from the days when northern coal ships tied up at the wharves.

Tyburn Brook flows past Soho, Mayfair, and the gallows of Tyburn Tree, its position marked on a traffic island. In Mayfair a narrow Tyburn channel replete with goldfish flows through the basement of an antique market (Fig. 21.4B), where the river was rediscovered during renovations.[13] The former channel runs through fashionable Berkeley Square where the magnificent plane trees dipped their young roots in an eighteenth-century water meadow. In 1862 a traveller 'voyaged' underground along the Tyburn sewer and, below Buckingham Palace, doffed his cap and sang the National Anthem. 'Well, you can take it from me', a sewerman near the palace said to author Eric Newby, 'that what comes down hasn't got "By Appointment" on it.'[14]

The Westbourne was mostly underground by the mid-nineteenth century. As *Household Words* noted,

the river was carefully tended in the Serpentine boating lake but continued as a foul sewer through the slums of Chelsea.[15] At Sloane Square tube station, the river flows in a metal culvert over the tracks.

And there is the Thames itself. In *The Pickwick Papers*, Dickens caught the excitement of the great river at dawn, the 'glistening water tinted with the light of the morning's sun and stirring with all the bustling preparations for business and pleasure that the river presented at that early hour'.[16] Flowing in from the south is the Neckinger, where in *Oliver Twist*, murderer Bill Sikes fell from a roof top to a hideous death, entangled in a noose of rope. As the ebb tide exposes the seawall, culverts with dripping pipes come into view, marking the confluences of the lost rivers with the Thames. In the shadows behind Blackfriars Bridge (Fig. 21.3), a grating marks the final trickle of the Fleet.

On the Embankment that he constructed is a monument to Joseph Bazalgette, a balding figure with a Victorian moustache and sideburns. Above is the epitaph *Flumini vincula posuit*—'He chained the river'.

CHAPTER 22

RESTORED RIVERS

Icelandic farmers blow up a dam

In August 1970, an explosion resonated around the hills at Lake Mývatn in Iceland, where the Eurasian and North American plates are pulling apart with lava eruptions that light up farmhouse kitchens kilometres away.[1] Fed by mineral springs, Lake Mývatn is rich in algae, plants, and insects that attract more species of nesting duck than anywhere else on Earth. From the lake the Laxá, a salmon river, runs through a narrow gorge where dams and power stations were built in the 1930s (Fig. 22.1). But new plans in 1964 called for a dam that would block the lake exit and raise its level, increasing the head of water for an additional power station on the river.

The ancestors of the Lake Mývatn inhabitants included the Viking Víga-Skútu (Killer-Skuta), who revenged the killing of his lawmaker father and for a while survived assassination attempts from men considered 'absolutely worthless'. He was eventually killed—'a wise man and a great hero', commented the Icelandic sagas, 'but not everyone thought him to be a reasonable man'.[2] Killer-Skuta's descendants had lived for centuries by farming, fishing and gathering eggs, traditionally leaving some in the nest. They were closely allied to the natural world, and they felt that the lake and surrounding land should not be sacrificed for a power dam. But they were few and nature had no vote. The

Figure 22.1 Laxá River in northern Iceland, draining to the ocean from Lake Mývatn.

Figure 22.2 Gullfoss Waterfall, Iceland, a major tourist site, which was narrowly saved from a dam and hydro development in the 1920s.

farmers drove to a nearby town with a sign 'Keep Laxá from harm', but the hydro developers made it known that land confiscation would follow if there was trouble.

A message to meet secretly at the dam was passed by word of mouth at a funeral, with no use of phones that could be tapped. The assembled crowd broke down the earth wall of the dam, but to their dismay found an inner concrete wall that would have to be taken out. It chanced that the hydro company had stored explosives in nearby caves to shift ice jams. One of the farmers knew how to use dynamite and was entertained by the prospect of destroying the dam with the company's own explosives. No one had blown up a dam before, but the group sang patriotic songs and the Internationale as the charges were fired. A huge explosion set the river flowing. 'You could hear Laxá calling, "I'm free, I'm free,"' said one of the farmers.

Police and powerplant officials searched for the ringleaders, but the community closed ranks. Some 60 individuals were charged as conspirators and others signed on, piqued that they had not been charged for what they considered an honour. But the company had acted illegally at many points, and the charges were eventually dropped. Few knew who had blown the dam until, ill and in hospital, the last surviving organizer confessed to moviemaker Grímur Hákonarson, who produced the 2013 documentary *Hvellur*. Lake Mývatn and the Laxá River are now formally protected.

Protest against dams is not new in Iceland. Gullfoss waterfall (Fig. 22.2), a major tourist attraction, was nearly lost to a hydro development in the 1920s. Sigríður Tómasdóttir travelled frequently to government offices on foot or horseback to argue for the waterfall, threatening to throw herself in if the dam went ahead. In the end, the plan was dropped. But the Kárahnjúkar Dam, nearly 200 m high, was constructed to power an aluminium plant and has been a source of great controversy. In the words of a farmer near Lake Mývatn, 'Visit the dam and see how we are capable of altering the landscape.'

Cleaning up the Hudson River

A group of commercial and recreational fishermen kick-started the cleanup of the Hudson River near New York and played a major role in North America's environmental movement.[3] After the Second World War, heavy industry, municipal works, and government institutions poured toxic chemicals into the river, which at times was in danger of catching fire, as it had done in 1921. Among those who pressed for river cleanup were fishermen Robert H. Boyle and John Cronin, and lawyer Robert F. Kennedy Jr, as documented in *The Riverkeepers*.

Early protests focused around a planned powerhouse at Storm King Mountain, a centrepoint of early American painting and a crucial spawning site for striped bass. As Boyle knew, the fish would be sucked up in huge numbers, because stacks of dead fish had been photographed at another powerhouse nearby. Boyle tracked down the photographer, who told him that officials had confiscated his original photos, returning later for the duplicates. Digging a set of photos from a drawer, the photographer said, 'But they never asked about triplicates.'[4] The photos provided evidence in a landmark legal decision that halted development and recognized the beauty and history of sites as legal evidence.

The fishermen began to organize. Their members included an airline pilot, a prison guard, an orthodontist, and a gravedigger who became the first head of what would become the *Riverkeeper* programme of the Hudson River Fishermen's Association. One member was a janitor who had complained fruitlessly to company managers and government officials about work-site pollution. After one call to officials, he observed them joking with the managers. A veteran of the war in Europe, he joined the fishermen, commenting, 'If Patton didn't get me killed nobody could.'[5] Local resident Pete Seeger, who had immortalized the river in his songs *Old Father Hudson* and *My Dirty Stream*, used a boat as an environmental classroom, and Cronin became involved while working with Seeger.

The fishermen and their allies pursued river polluters relentlessly, backed up by law students who won cases against corporate lawyers. They carried fishing rods to avoid suspicion, and they challenged government bodies who found it easier to shut down a fishery than to confront a polluter. They crawled up pipes, descended on ropes into culverts, donned scuba gear to gather dumped material, and watched for illegal flows down 'sneak pipes'. The river watchers could tell what colour cars were being painted in nearby factories that day, and they found blood and formaldehyde draining into the river from funeral homes. Tankers offloaded Caribbean oil and filled up with water for the home refineries after discharging their oily waste into the river. Gun clubs that shot at targets over water polluted the river with hundreds of tons of lead shot each year. Legal cases continue, but the Hudson River has been cleaned up substantially as a result of the Riverkeepers.

Daylighting Saw Mill River on the Hudson

Diving into New York's Grand Central Terminal with constellations painted on the ceiling, the traveller leaves the colossal human enterprise behind—the blaring rush-hour traffic, the towering skyscrapers. The train runs north along the Hudson in the gathering dusk and out to open country. Yonkers Station opens directly onto Saw Mill River, or Nepperhan as Native Americans called the Hudson tributary. Henry Hudson sailed past the river mouth, and the town was named for a young gentleman or *Jonkheer* who obtained an early land grant. The creek was enclosed in a culvert in the 1920s and paved over for a parking lot. But in 2011 the culvert was removed, and Saw Mill River was daylighted, seeing the sky for the first time in nearly a century (Fig. 22.3). The rushing water plunges over rapids reconstructed through the downtown, and a walkway with benches

Figure 22.3 Saw Mill Creek at Yonkers upstream from New York. From the 1920s, the creek ran to the Hudson River through a culvert but was 'daylighted' in 2011.

borders the creek, where bushes are flowering in an early spring. A fish ladder allows eels and alewife to surmount the rapids. Business has improved, says the owner of a nearby pizzeria, although the reconstruction was a difficult time. As an afterthought he adds that it is lovely to have the river back.

Children love the Sackville River

The classroom at Hammonds Plains school in Nova Scotia is a happy, messy place. There is the usual paraphernalia of early education—amazingly small desks and chairs and self-crayoned portraits with remarkable self-insight. The 25 students in Grades 1 and 2 are named on the door inside outlines of fish.

On a table stands a large fish tank, dripping water onto the floor. Two children have the daily task of checking the water temperature, which must be kept at 4°C. On the floor of the tank are hundreds of trout eggs, rolling gently in the flow, provided by the Fish Friends programme for the Sackville River Association.

The children are watching a video about the life cycle of the salmon. One girl is sitting on the floor and another is draped over a desk, leaping up to ask questions. On the screen a biologist holds a female salmon that is releasing eggs into a pail, and a male salmon is squirting sperm onto the eggs. Some of the children have covered their eyes but from somewhere an insistent voice is calling, 'It's not gross, it's not gross.' Salmon are leaping, and one boy is on the floor imitating a wriggling fish: he has a tank at home, he knows.

Petrie dishes on the desks have a trout egg and a floating ice cube to cool the distilled water. 'I've got trees on my arm,' says one student, studying his skin with a magnifying glass. The eggs are measured with a ruler and their contents are drawn in a large circle—two eyes, a curved red blood vessel, and a dark line for the growing kidney. The eggs are returned to the tank. 'You now know more than your parents,' says the Fish Friends leader. 'My dad doesn't know anything,' says a girl.

Six weeks later the tank contains a mass of tiny fish. Their mouth movements are visible in the petrie dishes, and the class copies the sucking motions. 'I saw the gills!' shouts someone. The fish will soon develop a digestive system, so the tank will need toilet paper, says the Fish Friends leader. Two girls have named their fish Goldie and Diamond. Some fish have two heads. 'Freak out!' a

boy mutters, but the children seem to take this in stride. The Fish Friends leader uses a brush to show how a fish tail brushes gravel aside to create a hollow or *redd* for the eggs, from which the newly hatched fish must wriggle out. A student wriggles out from under a chair without using arms or legs. Many chairs are knocked over.

It is a spring day on the Sackville River, and a fish ladder of small concrete pools connected by angled channels allows the fish to surmount a waterfall. The class gathers near the top of the ladder where 20 large fish are swimming in a chute, trapped by a gate system for the occasion. With a net the Fish Friends leader catches a fish and places it in the hands of a small girl (Fig. 22.4A), who holds it for a moment before returning it to the chute. Most of the fish are gaspereau or alewife, once so abundant here that settlers could almost cross the river on their backs. They are making

Figure 22.4A Children at the Sackville River, Nova Scotia, releasing fish captured at a fish ladder.

Figure 22.4B Children releasing fish fry into the Sackville River.

a comeback, but only a few salmon have returned. The children wipe off the fish scales onto their clothes. At a quiet stretch of the river, the students release more than a hundred fry from the tank (Fig. 22.4B). Keep in touch with Goldie and Diamond on Facebook, says the Fish Friends leader.

With a population of less than a million, Nova Scotia has about 80 river conservation groups. Tens of thousands of people—perhaps more than ten per cent of the population—are keenly interested in rivers, with some 60,000 fishing licences issued each year for trout and 3000 for salmon. There are still more than 500 dams in the southern part of the province, many of them orphans without oversight and liable to collapse,[6] and only a small proportion have fish passages. Collectively they have excluded salmon from more than 4000 river kilometres. A century ago, millponds along the Sackville River were clogged with sawdust from sawmills, but the dams are slowly coming down. An additional problem is the 50,000 river culverts that run under the province's roads, essentially dams with a small hole.

Back at school one girl has kept a souvenir—a large fish scale. 'It cheers me up,' said Pete Seeger. 'You can't look at those young faces and say that there's no hope'.[7]

Taking out the dams

Large dams continue to be built and waves of toxic waste sweep down channels. Irrigation takeout dries up rivers and deltas, while freshwater organisms fight for their lives. And hundreds of millions of people in the path of river development are dispossessed, disadvantaged, and diseased.

But there are signs of hope. Construction of some large dams has been cancelled or put on hold on the Amazon in Brazil and Peru, the Salween in China, the Irrawaddy in Myanmar, and in Australia, Chile, Iran, Lebanon, Portugal, and Thailand—temporary victories against a tide of permanent defeats. There is uncertainty over the huge Inga III dam at Inga Falls on the Congo River, slated to provide almost twice the hydropower of the Three Gorges Dam. A Waterkeeper Alliance based on the Hudson Riverkeeper model has more than 300 organizations worldwide, and there are many Water Protector organizations. Rivers have been daylighted in many cities. Parts of London's lost rivers may see the

light again, and a brewery has immortalized the rivers by using their names for varieties of beer.

Dams are being removed on many smaller rivers,[8] the stuff of fiction when *The Monkey Wrench Gang* was published. More than 5000 small dams, weirs, and culverts have been removed in Europe since the late 1990s, and by 2015 more than 1000 US dams had been taken down as arguments around indigenous claims, fisheries, and endangered species gained traction. Despite huge expenditures on hatcheries and fish passages, there is growing acceptance that fish populations can only recover if dams are removed.

The 2012 demolition of the century-old Elwha and Glines Canyon dams in Washington State cost more than $300 million and was the largest such project in American history (Fig. 22.5). Within the first free-flowing season, thousands of salmon returned to the upper Elwha, and marine nutrients—an important part of river ecology—reappeared in the ecosystem as the salmon spawned and died. The removal of two downstream dams on the Penobscot River in Maine, among a group of Atlantic coastal rivers that once hosted tens of millions of fish, opened some 1600 upriver kilometres for salmon and the former downriver habitat for sturgeon and striped bass. By allowing increased hydropower generation at other dams, the cost to companies and consumers was minimized.

Faced with a choice between hydropower and salmon on western American rivers, priority was given automatically to hydropower, as Aldo Leopold pointed out.[9] Debate has focused on the big dams on the Snake River, which formerly had enormous salmon runs. And the Glen Canyon Dam has come under scrutiny as Colorado flow declines through climate change. The enormous losses through evaporation and seepage from Reservoir Powell have been publicly discussed (they approximate the water supply of Los Angeles), but for now the dam remains.[10]

'You are one of us'

The Whanganui River rises in the volcanic uplands of New Zealand and flows to the Tasman Sea. Many indigenous communities lived on the river before Europeans arrived. In 2017, a court accorded the river the same legal status as a person, settling a case that had been

Figure 22.5 Remains of Glines Canyon Dam on the Elwha River in Washington State, removed in 2014 in the largest dam removal to date. Shutterstock©JDaracunas.

unresolved since the nineteenth century, and the river was considered a living whole from the mountains to the sea.[11] New Zealand's indigenous communities had often considered rivers as people—'I am the river and the river is me,' was a common saying. The legal ruling came with funding to improve the health of the river, which can be represented in law by guardians appointed by government and the community. Might the river now vote? people joke. Could it buy a few beers, depending on its age?

Citing the Whanganui ruling, a court in India declared that the Ganga and Yamuna rivers are people with cultural value since Vedic times, the first non-human entity in India to be granted human legal rights.[12] As a newspaper commented, they are among the country's most downtrodden people, with more than half the Ganga flow diverted for irrigation. In future,

polluting or damaging the rivers might be considered the legal equivalent of assault or murder.

Many societies have exercised the 'right' to engineer the Earth for our benefit as rivers are engineered for a dam,[13] but this view is changing. As prairie writer Trevor Herriot notes in *River in a Dry Land*:

> we have done much to reduce our conversation with the land to a monologue of demands and plunder, and although we do not have the myth-mind of the original listeners, we do have other faculties—the will, imagination, thought, emotion, and memory of modern man. With these we can again listen and respond to a mountain or a river.[14]

Recognizing rivers as givers of life, some societies have paid them the ultimate compliment: 'You are one of us.'

EPILOGUE

A journey of being

And there is another river, a river of the heart. For as long as I have studied geology, I have wanted to know about rivers. Following a host of scientists, I have explored how rivers cut their courses and build up strata, how they crossed supercontinents and responded to the breaking of continents and the rise of mountains. My journey has taken me back to a time so remote that we can only surmise that rivers existed, and I have observed ancestral rivers fossilized as rock. Earth's river journey spans four billion years with perhaps the same to come, and although Earth is for now the River Planet, our rivers have kin on ancestral Mars, on modern Titan, and almost certainly on alien worlds far out in Space.

My journey has been one of *knowing*. The Earth's ancient rivers have become so real to me that I can *feel* their existence, as Richard Jefferies felt the existence of the person buried in the tumulus, seemingly alive only seconds before. But my final journey is one of *being*. It began in childhood when I wanted to be in the flow of the river, building the dams of stones and moss that an amused channel tolerated for a while and then swept away.

As Edward Abbey knew, rivers are implacably indifferent to us—they owe us nothing and need nothing from us. But I cannot be indifferent to the rivers. And metaphor is not enough to explore the connection that I seek. We cannot step into the same river twice, said Greek philosopher Heraclitus. The river is everywhere at once and exists only in the present, said Hesse's ferryman Vasudeva. German rivers in winter reflect the destitute human heart, wrote Wilhelm Müller. But Humboldt knew that it was a privilege to be part of the life of rivers, if only for a moment. So did the Vedic writers on the Saraswati River, Tagore on the rivers of the Indian subcontinent, Li Bai and Du Fu along the rivers of China, and W.B. Yeats catching pike in the rivers of Ireland.

One day in the wooded valley of the Barle (Fig. ii.i), I sat for a long while watching the current rush over rock

Figure ii.i The River Barle on Exmoor in Devon, UK.

ledges. Oak logs carpeted the riverbed under the dark water, where a ripple marked a trout catching insects in the age-old exchange between water and land. Water trickled into the channel from a carpet of moss and swallows drank from the river in flight. There was an acceptable human presence, too—a prehistoric stone-slab bridge and the ford where medieval packhorses crossed. It seemed to me a lifetime's work to understand even one small river.

Flowing with the Barle in my memory were many rivers, living and long gone, ancestors or fellow travellers with the Barle and myself. I thought of the Saskatchewan River in the forests of Canada, Cooper Creek in the Australian desert, and the little Sackville River swollen with snowmelt. And out of the long history of the Earth's water cycle came the Torridonian rivers of Scotland and the Permian dryland rivers of Texas. I know these long-dead rivers turned to stone more intimately than I know any modern river, the graveyard markers of a former age. I know the ripples on their sand banks, the mud cracks where they dried up, the footprints of reptiles that forded the creeks.

Colin Fletcher, the first person known to have walked through the Grand Canyon, felt the rhythm and cadence of the rock, as real to him as a ticking clock and far more real than the beat of civilization on the canyon rim. We may be newcomers on Earth, he concluded, but we are exciting animals who owe their existence to the preceding eons. Despite all, we can be part of it.[1]

On the Chalk hills above the Thames, Richard Jefferies caught this vision a century earlier:

> It is eternity now. I am in the midst of it. It is about me in the sunshine; I am in it, as the butterfly floats in the light-laden air. Nothing has to come; it is now. Now is eternity; now is the immortal life. Here this moment, by this tumulus, on earth, now; I exist in it.[2]

Repaying our debt

A path winds up to Pinkworthy Pond on Exmoor through meadow grasses bright with dandelion, buttercup, and bog asphodel. Butterflies tumble across the valley and the infant Barle emerges below the overgrown earth dam between wind-sculpted beech hedgerows. A group of mountain bikers toils up the ascent—'Vive le peloton!' But only the hum of bees and the click of grasshoppers disturb the silence of the pond. A small plane drones overhead as its forerunner did fifty years before when I visited the pond as a teenager. I eat an orange in memory of Farmer Gammon.

On the forbidding landscape of the nearby Chain Barrows are the tumuli where prehistoric people buried their dead, along with a standing stone of sandstone with cross-beds that mark the flow of a long-extinct Devonian river. Oak and hazel once covered the Chain Barrows until Neolithic peoples cut down the forests. This is a landscape of our making, but springs still trickle from the bogs to source the Barle.

From what seems like the top of the world I look out over the sweep of geological time. To the east are the Cretaceous chalklands with their springs and dry valleys, and to the west lies the Atlantic where Pangea rifted apart, drawing the rivers to a widening ocean. To the south the Barle and the Exe once joined the Channel River where Stone Age peoples hunted until the great river was drowned.

The haze is clearing over the hills of Dartmoor to the south. There, high on the moorland, a hawthorn tree stands among ferns and heather in the headwaters of the River Teign. With his ashes buried among the hawthorn roots, my father is repaying his debt to the Earth as he would have wished.

GLOSSARY

A

Anastomosing: River with many stable channels that branch and rejoin, cut into floodplain sediment

Anticline: An elongate, convex-upward fold or arch in the strata

Avulsion: Temporary or permanent switching of a channel into a new course

B

Bank: Sloping margin of a river channel, with the top usually well above the water level

Bar: Mass of sediment between minor channels in a river or attached to the bank

Barbed Tributary: A stream that joins the mainstem channel with a bend that points upstream at the junction, commonly indicating that the mainstem channel has reversed its flow

Base Level: The level of a sea or lake, the deepest level to which a river can usually cut down

Basin: Topographically low area of the Earth's crust where sediment accumulates, often to thicknesses of kilometres through geological time

Braided: River with shallow and unstable channels between bars at low flow, resembling a braid of hair

C

Canyon: Deep, narrow valley commonly cut through a plateau

Channel: Water flowing above a bed of sediment and between banks

Clay: Sediment particles less than 4 microns in diameter

Conglomerate: Bed of gravel turned to rock through geological time

Cordillera: Elongate belt of mountain ranges and valleys along the western margin of North and South America

Craton: Stable part of the Earth's topmost layer or crust, commonly forming the old central area of a continent

Cross-beds: Downflow-dipping layers formed when sediment slides down the steep face of a dune or bar

D

Drainage basin: Area from which precipitation drains into a system of trunk and tributary river channels; separated from other drainage basins by watersheds or drainage divides; also known as catchment areas

Dune: Large ripple-like form generated by flowing water or wind

Dynamic topography: Raising and lowering of the Earth's crust, commonly as continents drift over mantle areas of variable temperature and density

E

Escape Tectonics: Lateral motion of large areas of the Earth's crust along faults, caused by an indenting continent, for example along the eastward extension of the Himalayan range

Eustasy: Global changes in sea level, caused by variation in the volume and temperature of sea water, or by large changes in bathymetry

F

Fault: A fracture or group of fractures along which rocks have moved relative to each other

Floodplain: Commonly vegetated area that borders a river and is covered by water and sediment during floods

Fold: A bend in the strata due to deformation, with upward and downward deflections (anticlines and synclines)

G

Gondwanaland: A formerly united group of continents in the southern hemisphere, including South America, Africa, India, Australia, and Antarctica

Graded River: A river that has achieved a balance between its ability to transport sediment and the amount of sediment supplied to it

Gravel: Rock fragments more than 2 mm in diameter, including granules, pebbles, cobbles, and boulders

Grenville Mountains: Mountain range once of Himalayan scale, generated by continental collisions that formed the Rodinia supercontinent about one billion years ago, eroded down and preserved along the eastern side of North America

Groundwater: Water in the spaces of a sediment or rock, especially below the water table

I

Imbrication: Gravel particles aligned with or perpendicular to the flow, commonly with flat surfaces dipping upstream

Incision: Downcutting of a river channel into sediment or rock below the riverbed

Indentor: Continent that has been forced into a larger continent during collision, causing strong deformation, for example India moving northwards into Asia

Isostasy: Gravitational balance that links the elevation of the crust to its density and thickness, as it 'floats' on the mantle

K

Karst landscape: A landscape with sinkholes, caves, and underground drainage systems, formed where relatively soluble limestone, dolomite, and gypsum are partly dissolved

Kimberlite An igneous rock that may contain diamonds, named for Kimberley in South Africa; the rock occupies volcanic pipes that brought magma to the surface from deep in the Earth's mantle, where the diamonds formed under high pressure

Knickpoint: Site where the river level changes over a short distance as a waterfall, rapids, or cataract

L

Levee: Sediment ridge alongside a channel, where floods lay down sediment, may be natural or artificial (human-made)

Limestone: Rock made of carbonate minerals (usually calcite, $CaCO_3$) and commonly formed in tropical seas

Loess: Wind-blown sediment, typically made of silt grains

M, N

Meandering: River with sinuous bends or meanders in a single channel; erosion and deposition on a point bar causes the channel to migrate across the floodplain

Mesolithic: Stone Age period between the Paleolithic and Neolithic, about 20,000 to 5000 years ago, with stone tools and locally pottery but predating organized agriculture

Mud and Mudstone: Sediment with a mixture of clay and silt, turned to mudstone rock

Neolithic: Stone Age period from the beginning of crop domestication and farming about 10,800 years ago, predating the use of metal tools

P

Paleolithic: Stone Age period from the oldest known use of stone tools by hominins to the start of the Mesolithic period

Pangea: Supercontinent that assembled about 300 million years ago in the Carboniferous and Permian, breaking up about 200 million years ago to form the modern continents

Peneplain (planation surface): Nearly flat surface tens to hundreds of kilometres wide formed by prolonged landscape weathering and river erosion

Plate: Areas of the Earth's lithosphere, comprising the outer crust (denser oceanic crust and less dense continental crust) and the underlying upper mantle

Plume: Hot magma rising from deep in the Earth's mantle, elevating the surface rocks and locally causing continents and ocean floors to rift and fragment

Point bar: Sediment laid down on a sloping surface in the bend of a meandering river

Power canal: Artificial water course diverted at a dam, directing much of the river flow past turbines for hydropower generation

R

Regolith: Fragmented and unconsolidated rock material that covers much of the land surface, including soil and glacial debris

Rift: During the tectonic stretching of the crust, an elongate depression forms, bounded by major faults that have lowered the central area; bordered by elevated rift flanks

Ripple: Small sediment wave formed by flowing water or wind

River: Moving body of water that flows from upland headwaters to a lake, ocean or another river, usually confined to a channel; small rivers are commonly termed brooks, creeks, and streams; rivers in dryland areas with periods of strong flow and minimal or no flow are termed washes

Rodinia: Supercontinent that formed about one billion years ago, breaking up about 750 million years ago

S

Sand and Sandstone: Grains of rock from 63 microns to 2 mm in diameter, mainly the mineral quartz (SiO_2), turned to sandstone rock

Sediment: Loose rock fragments and mineral grains of gravel, sand, silt, and clay

Shield: Large area of a craton with old rocks, typically with a gently convex-up surface

Silt: Grains of rock from 4 to 63 microns in diameter, mainly the mineral quartz; often wind-blown to accumulate as loess

Strata: Layers of sediment or rock, originally horizontal but commonly tilted through geological time; often referred to as beds

Syncline: An elongate, concave-upward fold or trough in the rock strata

Syntaxis: A prominent bend in a mountain belt, formed by rotation and compression around an indenting continent

T

Terrace: Layer of river sediment that marks a former floodplain and river level, abandoned after the channel cut down; numerous terraces with a deep valley form a staircase; sediment laid down on a surface cut into bedrock forms a strath terrace

Tethys Ocean: An ocean that once occupied the region of the Alpine–Himalayan mountain chain, largely obliterated by continental collisions

Thrust Sheet: Large mass of crustal rock that has moved laterally for tens to hundreds of kilometres above a gently dipping thrust fault

U

Unconformity: A surface within a body of rock or sediment that marks a substantial gap in time; the rock below was eroded before deposition of younger sediment above

Uniformitarianism: Principle that processes have acted through geological time much as they do today, often represented as 'the present is the key to the past'

V, W

Valley: River course cut into older sediment or rock, deeper and wider than the river that it contains

Water table: Surface of the groundwater, visible in a well

Weathering: Breaking down or dissolving of rocks and minerals at the Earth's surface through physical, chemical, and biochemical action

FURTHER READING

Chapter 1

Bhowmik, N., Richardson, E.V. and Julien, P.Y. (2008) Daryl B. Simons—hydraulic engineer, researcher, and educator. *Journal of Hydraulic Engineering* **134**, 287–294.

Braudrick, C.A. et al. (2009) Experimental evidence for the conditions necessary to sustain meandering in coarse-bedded rivers. *PNAS* **106**, 16936–16941.

Gibling, M.R., Nanson, G.C. and Maroulis, J.C. (1998) Anastomosing river sedimentation in the Channel Country of central Australia. *Sedimentology* **45**, 595–619.

Horn, J.D., Fielding, C.R. and Joeckel, R.M. (2012) Revision of Platte River alluvial facies model through observations of extant channels and barforms, and subsurface alluvial valley fills. *Journal of Sedimentary Research* **82**, 72–91.

Moody, J.A., Meade, R.H. and Jones, D.R. (2003) *Lewis and Clark's observations and measurements of geomorphology and hydrology, and changes with time*. United States Geological Survey, Circular **1246**.

Repcheck, J. (2009) *The Man Who Found Time: James Hutton and the Discovery of the Earth's Antiquity*. New York: Basic Books.

Chapter 2

Balme, M.R., et al. (2020) Aram Dorsum: An extensive mid-Noachian age fluvial depositional system in Arabia Terra, Mars. *Journal of Geophysical Research Planets* **125**, e2019JE006244.

Eriksson, K.A. and Wilde, S.A. (2010) Palaeoenvironmental analysis of Archaean siliciclastic sedimentary rocks in the west-central Jack Hills belt, Western Australia with new constraints on ages and correlations. *Journal of the Geological Society, London* **167**, 827–840.

Grotzinger, J.P., et al. (2015) Deposition, exhumation, and paleoclimate of an ancient lake deposit, Gale crater, Mars. *Science* **350**, aac7575-1 to 12.

Hadding, A. (1940) We and the world outside. Geological aspects of the problem of meteorites. *Kungliga Fysiografiska Sällskapets i Lund Förhandlingar* **10**, 37–51.

Rainbird, R.H. and Young, G.M. (2009) Colossal rivers, massive mountains and supercontinents. *Earth* **54**, 52–61.

Reimink, J.R., et al. (2014) Earth's earliest evolved crust generated in an Iceland-like setting. *Nature Geoscience* **7**, 529–533.

Rosenthal, E. (1970) *Gold! Gold! Gold! The Johannesburg Gold Rush*. London: Macmillan.

Chapter 3

Calder, J.H. (2006) 'Coal Age Galapagos': Joggins and the lions of nineteenth century geology. *Atlantic Geology* **42**, 37–51.

Gibling, M.R. and Davies, N.S. (2012) Palaeozoic landscapes shaped by plant evolution. *Nature Geoscience* **5**, 99–105.

Lovelock, J. (1995) *The Ages of Gaia, A Biography of our Living Earth*. New York: W.W. Norton.

Simon, S.S.T. and Gibling, M.R. (2017) Fine-grained meandering systems of the Lower Permian Clear Fork Formation of north-central Texas, USA: Lateral and oblique accretion on an arid plain. *Sedimentology* **64**, 714–746.

Stein, W.E., et al. (2012) Surprisingly complex community discovered in the mid-Devonian fossil forest at Gilboa. *Nature* **483**, 78–81.

Wohl, E. (2005) Disconnected rivers: Human impacts to rivers in the United States. In J. Ehlen, W.C. Haneberg and R.A. Larson (eds) *Humans as geologic agents*. Boulder: Geological Society of America Reviews in Engineering Geology **XVI**, 19–34.

Chapter 4

Leake, B.E. (2011) *The Life and Work of Professor J.W. Gregory FRS (1864–1932): Geologist, Writer and Explorer*. Geological Society, London, Memoir **34**.

Goudie, A.S. (2005) The drainage of Africa since the Cretaceous. *Geomorphology* **67**, 437–456.

Macgregor, D.S. (2012) The development of the Nile drainage system: integration of onshore and offshore evidence. *Petroleum Geoscience* **18**, 417–431.

Madof, A.S., Bertoni, C. and Lofi, J. (2019) Discovery of vast fluvial deposits provides evidence for drawdown during the late Miocene Messinian salinity crisis. *Geology* **47**, 171–174.

Potter, P.E. and Hamblin, W.K. (2006) Big rivers worldwide. *Brigham Young University Geology Studies* **48**, 1–78.

Chapter 5

Clarkson, C., et al. (2017) Human occupation of northern Australia by 65,000 years ago. *Nature* **547**, 306–310.

Craw, D., et al. (2016) Rapid biological speciation driven by tectonic evolution in New Zealand. *Nature Geoscience* **9**, 140–144.

Gregory, J.W. (1906) *The Dead Heart of Australia*. London: John Murray.

Habeck-Fardy, A. and Nanson, G.C. (2014) Environmental character and history of the Lake Eyre Basin, one seventh of the Australian continent. *Earth-Science Reviews* **132**, 39–66.

Jones, J. (1959) *The Cradle of Erewhon: Samuel Butler in New Zealand*. Austin: University of Texas Press.

Miall, A.D. and Jones, B.G. (2003) Fluvial architecture of the Hawkesbury Sandstone (Triassic), near Sydney, Australia. *Journal of Sedimentary Research* **73**, 531–545.

Sugden, D. (2009) The Dry Valleys: An ancient and cold desert in Antarctica. In P. Migoń (ed.) *Geomorphological landscapes of the world*. Springer Science Business Media B.V.2010, 113–131.

Chapter 6

Brookfield, M.E. (1998) The evolution of the great river systems of southern Asia during the Cenozoic India–Asia collision: rivers draining southwards. *Geomorphology* **22**, 285–312.

Gregory, J.W. and Gregory, C.J. (1923) *To the Alps of Chinese Tibet*. London: Seeley, Service & Company.

Hu, X., et al. (2016) The timing of India–Asia collision onset— Facts, theories, controversies. *Earth-Science Reviews* **160** 264–299.

Oldham, R.D. (1907) The valleys of the Himalayas. *The Geographical Journal* **30**, 512–516.

Pan, B., et al. (2012) The approximate age of the planation surface and the incision of the Yellow River. *Palaeogeography, Palaeoclimatology, Palaeoecology* 356–357, 54–61.

Şengör, A.M.C. (2015) The founder of modern geology died 100 years ago: The scientific work and legacy of Eduard Suess. *Geoscience Canada* 42, 181–246.

Chapter 7

Amorosi, A., et al. (2008) Late Quaternary palaeoenvironmental evolution of the Adriatic coastal plain and the onset of Po River Delta. *Palaeogeography, Palaeoclimatology, Palaeoecology* 268, 80–90.

Bridgland, D.R., et al. (2006) The Palaeolithic occupation of Europe as revealed by evidence from the rivers: data from IGCP 449. *Journal of Quaternary Science* 21, 437–455.

Gábris, G. and Nádor, A. (2007) Long-term fluvial archives in Hungary: response of the Danube and Tisza rivers to tectonic movements and climatic changes during the Quaternary: a review and new synthesis. *Quaternary Science Reviews* 26, 2758–2782.

Herget, J., et al. (2015) The millennium flood of July 1342 revisited. *Catena* 130, 82–94.

Olariu, C., Krezsek, C. and Jipa, D.C. (2018) The Danube River inception: Evidence for a 4 Ma continental-scale river born from segmented ParaTethys basins. *Terra Nova* 30, 63–71.

Popov, S.V., et al. (2006) Late Miocene to Pliocene palaeogeography of the Paratethys and its relation to the Mediterranean. *Palaeogeography, Palaeoclimatology, Palaeoecology* 238, 91–106.

Trümpy, R. (2001) Why plate tectonics was not invented in the Alps. *International Journal of Earth Science* 90, 477–483.

Chapter 8

Albert, J.S., Lovejoy, N.R. and Crampton, W.G.R. (2006) Miocene tectonism and the separation of cis- and trans-Andean river basins: Evidence from Neotropical fishes. *Journal of South American Earth Sciences* 21, 14–27.

Figueiredo, J., et al. (2009) Late Miocene onset of the Amazon River and the Amazon deep-sea fan: Evidence from the Foz do Amazonas Basin. *Geology* 37, 619–622.

Latrubesse, E.M., et al. (2010) The Late Miocene paleogeography of the Amazon Basin and the evolution of the Amazon River system. *Earth-Science Reviews* 99, 99–124.

Potter, P.E. (1997) The Mesozoic and Cenozoic paleodrainage of South America: a natural history. *Journal of South American Earth Sciences* 10, 331–344.

Wulf, A. (2014) *The Invention of Nature, Alexander von Humboldt's New World*. New York: A.A. Knopf.

Chapter 9

Blum, M., and Pecha, M. (2014) Mid-Cretaceous to Paleocene North American drainage reorganization from detrital zircons. *Geology* 42, 607–610.

Galloway, W.E., Whiteaker, T.L. and Ganey-Curry, P. (2011) History of Cenozoic North American drainage basin evolution, sediment yield, and accumulation in the Gulf of Mexico basin. *Geosphere* 7, 938–973.

Poag, C.W., et al. (1990) Early Cretaceous shelf-edge deltas of the Baltimore Canyon Trough: Principal sources for sediment gravity deposits of the northern Hatteras Basin. *Geology* 18, 149–152.

Potter, P.E. (1978) Significance and origin of big rivers. *Journal of Geology* 86, 13–33.

Ranney, W. (2012) *Carving Grand Canyon*. Grand Canyon, Arizona: Grand Canyon Association.

Reidel, S.P. and Tolan, T.L. (2013) The late Cenozoic evolution of the Columbia River system in the Columbia River flood basalt province. In S.P. Reidel et al., (eds) *The Columbia River Flood Basalt Province*. Boulder: Geological Society of America, Special Publication 497, 201–230.

Chapter 10

Leckie, D.A. and Cheel, R.J. (1989) The Cypress Hills Formation (Upper Eocene to Miocene): a semi-arid braidplain deposit resulting from intrusive uplift. *Canadian Journal of Earth Sciences* 26, 1918–1931.

Chapter 11

Cordier, S., et al. (2017) Of ice and water: Quaternary fluvial response to glacial forcing. *Quaternary Science Reviews* 166, 57–73.

Duk-Rodkin, A., et al. (2001) Geologic evolution of the Yukon River: implications for placer gold. *Quaternary International* 82, 5–31.

Hidy, A.J., et al. (2013) A latest Pliocene age for the earliest and most extensive Cordilleran Ice Sheet in northwestern Canada. *Quaternary Science Reviews* 61, 77–84.

Imbrie, J. and Imbrie, K.P. (1979) *Ice Ages: Solving the Mystery*. Short Hills: Enslow Publishers.

Shugar, D.H., et al. (2017) River piracy and drainage basin reorganization led by climate-driven glacier retreat. *Nature Geoscience* 10, 370–375.

Simms, M.J. and Coxon, P. (2017) The pre-Quaternary landscape of Ireland. In P. Coxon, S. McCarron and F. Mitchell (eds) *Advances in Irish Quaternary Studies*. Paris: Atlantis Press.

Chapter 12

Baker, V.R. (1995) Joseph Thomas Pardee and the Spokane Flood Controversy. *GSA Today* 5, 169–173.

Carling, P.A. (1996) Morphology, sedimentology and palaeohydraulic significance of large gravel dunes: Altai Mountains, Siberia. *Sedimentology* 43, 647–664.

Clague, J.J., et al. (2003) Paleomagnetic and tephra evidence for tens of Missoula floods in southern Washington. *Geology* 31, 247–250.

Waitt, R.B. (2002) Great Holocene floods along Jökulsá á Fjöllum, north Iceland. In I.P. Martini, V.R. Baker and G. Garzón (eds) *Flood and Megaflood Processes and Deposits: Recent and Ancient Examples*. Oxford: International Association of Sedimentologists, Special Publication 32, 37–51.

Zuffa, G.G., et al. (2000) Turbidite megabeds in an oceanic rift valley recording jökulhlaups of Late Pleistocene glacial lakes of the western United States. *Journal of Geology* 108, 253–274.

Chapter 13

Alqahtani, F.A., et al. (2017) Controls on the geometry and evolution of humid-tropical fluvial systems: Insights from 3D seismic geomorphological analysis of the Malay Basin, Sunda Shelf, Southeast Asia. *Journal of Sedimentary Research* 87, 17–40.

Bridgland, D.R. (2013) John Lubbock's early contribution to the understanding of river terraces and their importance to geography, archaeology and earth science. *The Royal Society Journal of the History of Science*, doi:10.1098/rsnr.2013.0053.

Fyles, J.G. (1990) Beaufort Formation (Late Tertiary) as seen from Prince Patrick Island, Arctic Canada. *Arctic* 43, 393–403.

Gupta, S., et al. (2017) Two-stage opening of the Dover Strait and the origin of island Britain. *Nature Communications* 8, doi:10.1038/ncomms15101.

Hanebuth, T.J.J., et al. (2011) Formation and fate of sedimentary depocentres on Southeast Asia's Sunda Shelf over the past sea-level cycle and biogeographic implications. *Earth-Science Reviews* 104, 92–110.

Nunn, P.D. and Reid, N.J. (2016) Aboriginal memories of inundation of the Australian coast dating from more than 7000 years ago. *Australian Geographer* 47, 11–47.

Overeem, I., et al. (2001) The Late Cenozoic Eridanos delta system in the Southern North Sea Basin: a climate signal in sediment supply? *Basin Research* 13, 293–312.

Chapter 14

Arranz-Otaegui, A., et al. (2016) Regional diversity on the timing for the initial appearance of cereal cultivation and domestication in southwest Asia. *PNAS* **113**, 14001–14006.

Ashley, G.M., et al. (2010) Paleoenvironmental and paleoecological reconstruction of a freshwater oasis in savannah grassland at FLK North, Olduvai Gorge, Tanzania. *Quaternary Research* **74**, 333–343.

Butler, D.R. (2006) Human-induced changes in animal populations and distributions, and the subsequent effects on fluvial systems. *Geomorphology* **79**, 448–459.

Clement, C.R., et al. (2015) The domestication of Amazonia before European conquest. *Proceedings of the Royal Society B*, **282**, 20150813.

Fuller, D.Q., et al. (2010) Consilience of genetics and archaeobotany in the entangled history of rice. *Archaeological and Anthropological Sciences* **2**, 115–131.

Gibling, M.R. (2018) River systems and the Anthropocene: A late Pleistocene and Holocene timeline for human influence. *Quaternary* **1**, doi:10.3390/quat1030021.

Gibling, M.R., et al. (2008) Quaternary fluvial and eolian deposits on the Belan River, India: paleoclimatic setting of Paleolithic to Neolithic archeological sites over the past 85,000 years. *Quaternary Science Reviews* **27**, 391–410.

Hillel, D. (1994) *Rivers of Eden*. New York: Oxford University Press.

Mays, L.W. (2008) A very brief history of hydraulic technology during antiquity. *Environmental Fluid Mechanics* **8**, 471–484.

Roebroeks, W. and Villa, P. (2011) On the earliest evidence for habitual use of fire in Europe. *PNAS* **108**, 5209–5214.

Rosen, A.M. (2008) The impact of environmental change and human land use on alluvial valleys in the Loess Plateau of China during the Middle Holocene. *Geomorphology* **101**, 298–307.

Walter, R.C. and Merritts, D.J. (2008) Natural streams and the legacy of water-powered mills. *Science* **319**, 299–304.

Chapter 15

Danino, M. (2010) *The Lost River, On the Trail of the Sarasvati*. New Delhi: Penguin Books.

Singh, A., et al. (2017) Counter-intuitive influence of Himalayan river morphodynamics on Indus Civilization urban settlements. *Nature Communications* **8**, doi:10.1038/s41467-017-01643-9.

Chapter 16

Ball, P. (2017) *The Water Kingdom: A Secret History of China*. Chicago: The University of Chicago Press.

Dodgen, R.A. (2001) *Controlling the Dragon: Confucian Engineers and the Yellow River in Late Imperial China*. Honolulu: University of Hawai'i Press.

Dutch, S. I. (2009) The largest act of environmental warfare in history. *Environmental and Engineering Geoscience* **15**, 287–297.

Chapter 17

Dai Qing (1998) *The River Dragon Has Come!* Armonk: M.E. Sharpe.

Gilbert, G.K. (1917) *Hydraulic-mining debris in the Sierra Nevada*. United States Geological Survey, Professional Paper **105**.

Lehner, B., et al. (2011) High-resolution mapping of the world's reservoirs and dams for sustainable river-flow management. *Frontiers in Ecology and Environment* **9**, 494–502.

Lenders, H.J.R., et al. (2016) Historical rise of waterpower initiated the collapse of salmon stocks. *Scientific Reports* **6**, 29269, doi:10.1038/srep29269.

Macy, C. (2010) *Dams*. New York: W.W. Norton.

McCully, P. (1996) *Silenced Rivers*. London: Zed Books.

Vörösmarty, C.J., et al. (2010) Global threats to human water security and river biodiversity. *Nature* **467**, 555–561.

Wong, C.M., et al. (2007) *World's Top 10 Rivers at Risk*. Gland, Switzerland: WWF (World Wildlife Fund) International.

Worster, D. (2008) *A Passion for Nature; The Life of John Muir*. Oxford: Oxford University Press.

Chapter 18

Abbey, E. (1968) *Desert Solitaire*. New York: Touchstone.

Blair, T.C. (1987) Sedimentary processes, vertical stratification sequences, and geomorphology of the Roaring River alluvial fan, Rocky Mountain National Park. *Journal of Sedimentary Petrology* **57**, 1–18.

Dafalla, H. (1975) *The Nubian Exodus*. London: C. Hurst and Company.

Farmer, J. (1999) *Glen Canyon Dammed*. Tucson: The University of Arizona Press.

Leslie, J. (2005) *Deep Water*. New York: Picador.

Reisner, M. (1993) *Cadillac Desert, the American West and its Disappearing Water*. New York: Penguin Books.

Yi Si (1998) The world's most catastrophic dam failures. In Dai Qing (ed.) *The River Dragon Has Come!* Armonk: M.E. Sharpe, 25–38.

Chapter 19

Chodkiewicz, J.-L., and Brown, J. (1999) *First Nations and Hydroelectric Development in Northern Manitoba*. Winnipeg: The Centre for Rupert's Land Studies at The University of Winnipeg.

Smith, N.D., et al. (1998) The 1870s avulsion of the Saskatchewan River. *Canadian Journal of Earth Sciences* **35**, 453–466.

Smith, N.D., et al. (2016) Dam-induced and natural channel changes in the Saskatchewan River below the E.B. Campbell Dam, Canada. *Geomorphology* **269**, 186–202.

World Wildlife Fund (2009) *Canada's Rivers at Risk*, 17 pp.

Chapter 20

Brice, W.R. and Figueirôa, S.F.de M. (2003) Charles Frederick Hartt -- a pioneer of Brazilian Geology. *GSA Today* **March 2003**, 18–19.

Leopold, A. (1949) *A Sand County Almanac, and Sketches Here and There*. London: Oxford University Press.

Schmidt, M.J., et al. (2014) Dark earths and the human built landscape in Amazonia: a widespread pattern of anthrosol formation. *Journal of Archaeological Science* **42**, 152–165.

Stegner, W. (1954) *Beyond the Hundredth Meridian, John Wesley Powell and the Second Opening of the West*. Boston: Houghton Mifflin.

Worster, D. (1979) *Dust Bowl: The Southern Plains in the 1930s*. New York: Oxford University Press.

Chapter 21

Barton, N. (1962) *The Lost Rivers of London*. London: Historical Publications.

Halliday, S. (2009) *The Great Stink of London*. Stroud: The History Press.

Hill, A.B. (1955) Snow – An Appreciation. *Proceedings of the Royal Society of Medicine* **48**, 46–50.

Chapter 22

Cronin, J. and Kennedy, R.F.Jr. (1997) *The Riverkeepers*. New York: Scribner.

New Zealand declares a river a person. 2017, March, 25. The Economist. Retrieved on 19 December 2017 from https://www.economist.com/news/asia/21719409-odd-legal-status-intended-help-prevent-pollution-and-other-abuses-new-zealand-declares

O'Connor, J.E., Duda, J.J. and Grant, G.E. (2015) 1000 dams down and counting. *Science* **348**, 496–497.

The Laxá Farmers (2013). Movie directed by Grímur Hákonarson, Ground Control Productions.

CHAPTER ENDNOTES

A file listing full reference citations for the publications, may be downloaded from https://bit.ly/2PW1C1M

Prologue

1 Blackmore, 1869, pp. 13–14.
2 Menai, H. (Huw Owen Williams), 'geology is the infinite biography of God' (1920) The Geologist, *Through the Upcast Shaft.* London: Hodder and Stoughton, p. 30.
3 Greene, 2015, p. 65.
4 Jefferies, R. 'I wonder to myself' (1889) *Field and Hedgerow.* London: Longmans, Green and Company, p. 2.
5 Worster, D. 'Would you murder your own children?' (2008) *A Passion for Nature; The Life of John Muir.* Oxford: Oxford University Press, 383, citing reminiscence by Emily Bell, 13 October 1915, from John Muir Papers.

Chapter 1: Rivers and Geological Time

1 Duffin and Davidson, 2011; belemnites as thunderbolts from Miller, 1841, and personal statements to the author.
2 Cutler, 2003.
3 Richter, 1980, pp. 28–31.
4 Repcheck, 2009.
5 Playfair, J. 'The mind seemed to grow giddy' (1805) Biographical account of the late Dr. James Hutton, FRS, Edinburgh. *Proceedings of the Royal Society of Edinburgh* **5**, 73.
6 Torrens, 2001.
7 Repcheck, 2009.
8 England et al., 2007.
9 Folk, 1965.
10 Bhowmik et al., 2008.
11 Moody et al., 2003.
12 Tal and Paola, 2007.
13 Fisk, 1944.
14 McPhee, 1989.
15 Twain, M. 'One who knows the Mississippi' (1883) *Life on the Mississippi.*
16 Worster, 1979, p. 23.
17 Smith, 1970; Horn et al., 2012.
18 Nanson and Huang, 1999; Gibling et al., 1998.
19 Hovius 1996; DeCelles and Cavazza, 1999; Weissmann et al., 2015.
20 Seni, 1980; Goodwin and Diffendal, 1987; Heller et al., 2003.

Chapter 2: The First Drop of Rain on the Nascent Earth

1 Cameron, 2001; Amelin and Ireland, 2013.
2 Hjelmqvist, 1962.
3 Hadding A. 'An ash-covered earth' (1929) The first rains and their geological significance. *Geologiska Föreningen i Stockholm Förhandlingar* **51**, 21.

4 Hadding's work on meteorites from Corlin, 1939; Cross, 1947; Wickman and Uddenberg-Andersson, 1982.
5 Hadding, 1940.
6 Mangold et al., 2004.
7 Spaggiari et al., 2007; Eriksson and Wilde, 2010; Valley et al., 2014.
8 Harrison, 2009.
9 Battistuzzi et al., 2004; Driese et al., 2011; Wickramasinghe and Smith, 2014; Tashiro et al., 2017.
10 Follmann and Brownson, 2009.
11 Science told life started in hot water, 1931; Life, 1931.
12 Reimink et al., 2014.
13 Moorbath, 2005.
14 Eriksson et al., 2013.
15 Rosenthal, 1970; Wheatcroft, 1986.
16 Buck, 1983; Heinrich, 2015.
17 Naeraa et al., 2012; Bercovici and Ricard, 2014; Satkoski et al., 2017.
18 Bradley, 2011.
19 Li et al., 2008; Rivers et al., 2012.
20 Stewart, 2002; Ielpi and Ghinassi, 2015; Ielpi et al., 2016.
21 Callow et al., 2011; Strother et al., 2011.
22 Rainbird et al., 1997.
23 Rainbird and Young, 2009.
24 Oldroyd, 1990, p. 250.
25 See Hartmann, 2003 for a history of Mars exploration, geography, and geology.
26 Former Mars river systems from Mangold et al., 2004; Wood, 2006; Burr et al., 2009; Warner et al., 2010; Williams et al., 2013; Grotzinger et al., 2015; Baker et al., 2015; Balme et al., 2020.
27 Burr et al., 2013.
28 Heller, 2015.

Chapter 3: How Plants Bent and Split Rivers

1 Muir, 1997, The American Forests, p. 701.
2 Wohl, 2005.
3 Vegetation effects in rivers from Sedell et al., 1990; Parker, 1992; Wohl, 2013.
4 Early plant evolution and co-evolution with rivers from Greb and DiMichele, 2006; Davies and Gibling, 2010; Stein et al., 2012; Gibling and Davies, 2012; Gibling et al., 2014; Salamon et al., 2018.
5 Scottish geologists (Hugh Miller, Robert Dick, Charles Peach, and William Mackie) from Peach, 1879; Smiles, 1879; Obituary of Charles William Peach, 1886; Rosie, 1982; Trewin, 2004;

Anderson, 2005; displays at Caithness Horizons Museum, Thurso. Canadian geologists (William Logan, William Dawson, and Charles Lyell) from Calder, 2006, 2012; Dawson, 1855, 1858, 1859; Logan et al., 1863; Lyell, 1855; Torrens, 1999; Shipley, 2002; Falcon-Lang and Calder, 2005; Rygel and Shipley, 2005.

6 Miller, 1841, p. 9.

7 Smiles, S. 'How can I or any man' (1879) *Robert Dick, Baker, of Thurso: Geologist and Botanist*. New York: Harper & Brothers, pp. 309–10.

8 Smiles S. 'These are pertinent questions', (1879) ibid., p. 93.

9 *Prototaxites* from Dawson, 1859; Boyce et al., 2007.

10 Devonian rocks and fossils of Gaspé and northern New Brunswick from Shear et al., 1996; Gensel and Edwards, 2001; Davies et al., 2011; Kennedy et al., 2012; Kennedy et al., 2013.

11 Calder, 2006.

12 Rygel and Shipley, 2005.

13 Darwin, C. 'we have the plainest evidence' (1859) *On the Origin of Species*: London: John Murray.

14 Wilberforce, S. 'the rare land shell' (1860) Review of 'On the Origin of Species': *Quarterly Review 1860*, 244.

15 Joggins river systems from Falcon-Lang et al., 2004; Rygel and Gibling, 2006; Davies and Gibling, 2011; Ielpi et al., 2014, 2015; Prescott et al., 2014.

16 Leopold, 1994.

17 Lovelock, 1995.

18 Simon and Gibling, 2017.

19 Pangean assembly and rivers from Veevers, 2004; Potter and Hamblin, 2006; Gehrels et al., 2011; Seton et al., 2012.

20 Zambito and Benison, 2013.

Chapter 4: Breaking Pangea: The Ancestral Rivers of Africa

1 Greene, M.T. 'I'm going to have to pursue this' (2015) *A lfred Wegener*: Baltimore: Johns Hopkins University Press, p. 214.

2 Crane, 2003; Brotton, 2013, pp. 17–53.

3 Kious and Tilling, 1994.

4 Information about Alfred Wegener from Greene, 2015.

5 White, G. 'But this is making use of a violent piece of machinery' (1788–9) *The Natural History of Selborne*, letter XXIV.

6 Nield, 2007, pp. 65–67; Scott, 1913, pp. 340–341.

7 Cherry-Garrard, A. 'these same specimens' (1922) *The Worst Journey in the World*. London: Constable & Co.

8 Willis, 1944.

9 Mantle plumes and dynamic topography from Cox, 1989; Courtillot et al., 2003; Flament et al., 2013.

10 Potter, P.E. 'Water has always run downhill' (1997) The Mesozoic and Cenozoic paleodrainage of South America: a natural history. *Journal of South American Earth Sciences*, **10**, 331–344.

11 River responses to tectonic activity from Mann and Thomas, 1968; Potter, 1978, 1997; Potter and Hamblin, 2006; Ashworth and Lewin, 2012.

12 Landscape evolution of southern Africa from Cox, 1989; Moore and Blenkinsop, 2002; Goudie, 2005; Walford and White, 2005.

13 East African Rift System from Chorowicz, 2005; Corti, 2009; Moucha and Forte, 2011; Roberts et al., 2012.

14 African exploration from Moorehead, 1971, 1972; Hibbert, 1982; Van Orman, 1983.

15 Information about Jack Gregory from Leake, 2011.

16 Herodotus, The History; for the Nile, see Book 2, parts 19–28.

17 Nile exploration from Hibbert, 1982.

18 Nile River from Williams et al., 2003; Goudie, 2005; Gani et al., 2007; Gargani et al., 2010; Macgregor, 2012; Madof et al., 2019.

19 Niger and Senegal rivers from Burke et al., 1971; Makaske, 2001; Goudie, 2005; Potter and Hamblin, 2006; Macgregor, 2012; Bonne, 2014.

20 Hibbert, 1982, p. 69.

21 Joseph Conrad in Africa from Conrad, 1964; Jean-Aubry, 1964; Stape, 2007.

22 Conrad, J. 'resembling an immense snake' (1899) *The Heart of Darkness*.

23 Jean-Aubry, 1964, p. 100.

24 Hibbert, C. 'the river that swallows all others' (1982) *Africa Explored*. London: Allen Lane.

25 Conrad, J. 'Going up that river', (1899) ibid.

26 Congo River from Peters and O'Brien, 2001; Stankiewicz and de Wit, 2006; Runge, 2007; Harcourt and Wood, 2012; Macgregor, 2013; Garzanti et al., 2019.

27 Kingsley, M.H. 'One appalling corner' (1897) *Travels in West Africa*. London: MacMillan, pp. 169–174.

28 Wonham et al., 2000; Macgregor, 2013.

29 Livingstone, D. 'Have you smoke that sounds' (1857) *Missionary Travels and Researches in South Africa*. Information about David Livingstone from Hibbert, 1982.

30 Zambezi and Limpopo rivers from Salman and Abdula, 1995; Moore and Coterill, 2009; Moore et al., 2007; Braun et al., 2014.

31 Tigerfish from Moore et al., 2007; Goodier et al., 2011.

32 Diamonds in southern Africa from Kanfer, 1993; Waltham, 1997.

33 Kanfer, S. 'We have enough' (1993) *The Last Empire: De Beers, Diamonds, and The World*. New York: Farrar Straus Giroux, p. 34.

34 Orange River from Cox, 1989; Jacob et al., 1999; Rouby et al., 2009; Braun et al., 2014.

35 Future plate tectonics from Williams and Nield, 2007; Safonova and Maruyama, 2014; Scotese, 2014.

Chapter 5: Hot and Cold: The River Histories of Australia, New Zealand, and Antarctica

1 Gregory, 1906; Leake, 2011.

2 Leake, B.E. 'ink in the blood' (2011), p. 40.

3 Moorehead, 1963, p. 11, and exploration accounts.

4 Gregory, 1906, p. 95.

5 Gregory, 1906, p.78.

6 Australian rivers and El Niño from Finlayson and McMahon, 1988; Chiew et al., 1998.

7 Tectonic history of Australia from Pillans, 2007; DiCaprio et al., 2009; Quigley et al., 2010; Czarnota et al., 2014.

8 Miall and Jones, 2003.

9 Veevers et al., 2008.

10 Great Dividing Range and Shoalhaven River from Nott et al., 1996; Nott and Horton, 2000; Czarnota et al., 2014.

11 Channel Country rivers, Lake Eyre, and climatic history from Page and Nanson, 1996; Alley, 1998; Gibling et al., 1998; Habeck-Fardy and Nanson, 2014.

12 Murray-Darling River and south-coast canyons from Exon et al., 2005; Hill et al., 2005.

13 Twidale, 2009.

14 Australian megafauna, rock art, and dating from Roberts and Brook, 2010; Gunn et al., 2011; Mulvaney, 2013; Clarkson et al., 2017.

15 Discovery Expedition from Scott, 1905, p. 610-628.

16 Scott, R.F. 'as we ran the comparatively warm sand' (1905) *The Voyage of the 'Discovery'*. London: John Murray.

17 Taylor's discoveries from Taylor, 1958; Scott, 1913.

18 Cherry-Garrard, 'Old Griff on a sledge journey' (1922), ibid.

19 Sanderson, M. 'Damn nonsense!'(1988) *Griffith Taylor: Antarctic Scientist and Pioneer Geographer*: Ottawa: Carleton University Press, p. 63.

20 Dry Valleys and Transantarctic Mountains from David and Priestley, 1909, p. 91; Scott, 1913, pp. 329–334; Taylor, 1914; Sugden, 2009.

21 Butler, S. 'Property, marriage, the law' (1872) *Erewhon*. London: Trübner and Ballantyne.
22 Norris and Cooper, 2001.
23 Geological history of New Zealand from Youngson et al., 1998; Upton et al., 2004; Craw et al., 2016.
24 Samuel Butler in New Zealand from Jones, 1959.
25 Butler, S. 'The river-bed was here' (1872), ibid.
26 Butler, S. 'I am there now' (1872), ibid.

Chapter 6: Young and Restless: The Evolving Rivers of Asia
1 Muir, J. 'Not yet' (1915), *Travels in Alaska*. Boston: Houghton Mifflin.
2 Gibling et al., 1994.
3 Myrow et al., 2009.
4 Kipling, R. 'the town crier' (1894) *The Jungle Book*. London: Macmillan.
5 Information about Eduard Suess from Gregory, 1929; Şengör, 2014, 2015.
6 Suess, 1883–1909.
7 Greene, 2015, pp. 241–244 and 263.
8 Collision of India and Asia and monsoon history from Tapponnier et al., 2001; Yin, 2006; Clift et al., 2008; Hu et al., 2016.
9 Tagore, R. 'strata of human geology' (1931) The Religion of Man. The MacMillan Company.
10 Silk Roads from Christian, 2000; Frankopan, 2015.
11 Encyclopedia Britannica, 1984.
12 Great Trigonometrical Survey from Keay, 2000.
13 Secret surveying, European exploration, and the Great Game from Severin, 1976; Hopkirk, 1992.
14 Kipling, R. 'make pictures of roads' (1901) *Kim*. London: Macmillan & Co.
15 Pascoe, 1919; Fermor, 1949.
16 Longmuir, 2000.
17 Siwalik Group from Willis, 1993; Najman et al., 2009.
18 Oldham, 1907.
19 Wager, 1933, 1937.
20 Indus River from Brookfield, 1998; Abbasi and Friend, 2000; Clift et al., 2001; Clift, 2002; Clift and Blusztajn, 2005.
21 Ganga River from Brookfield, 1998; DeCelles et al., 1998; Yin, 2006.
22 Govin et al., 2018.
23 Brookfield, 2008.
24 Rivers of Peninsular India from Casshyap and Aslam, 1992; Sant and Karanth, 1993; Sahu et al., 2013; Nagendra and Reddy, 2017.
25 Gregory's expedition from Gregory and Gregory, 1923.
26 Barbour, 1936.
27 Gregory and Gregory, 'when the wave of the Himalaya' (1923), 176–7.
28 Willis, B.J. 'imprisoned by tremendous canyon walls' (1907) *Research in China*. Washington: Carnegie Institution of Washington, 334–335.
29 Rivers of E and SE Asia from Hallet and Molnar, 2001; Clark et al., 2004; Hall and Morley, 2004; Rüber et al., 2004; Hoang et al., 2009; Van der Beek et al., 2009; Richardson et al., 2010; Zheng et al., 2013; Alqahtani et al., 2015; Bracciali et al., 2015; Li et al., 2017; Govin et al., 2018; Fu et al., 2020.
30 Potter and Hamblin, 2006.
31 Ball, 2017, p. 14.
32 Loess Plateau from Sun et al., 2010; Stevens et al., 2013.
33 Slingerland and Smith, 2004.
34 Yellow River from Pan et al., 2012; Craddock et al., 2010; Wang et al., 2012; Zhang et al., 2014; Hu et al., 2017.
35 Information about Hermann Hesse from Freedman, 1978.
36 Hesse, H. 'Yes, it is a very beautiful river' (1922) *Siddartha*.
37 Information about Rabindranath Tagore from Kripalani, 1962.
38 Tagore, 1931, ibid.
39 Tagore, R. 'The heaven's river' (1910) *Gitanjali*. Poem 57.

Chapter 7: The Conflicted Rivers of Europe
1 Scott, W. 'I am now a deceased person' (1896) *The Downfall of Napoleon*. London: Blackie & Son, p. 12.
2 Eurite information from Serri et al., 2001; Pascucci et al., 2006; Dini et al., 2007.
3 Information about Nicolaus Steno from Cutler, 2003.
4 Trümpy, 2001.
5 Geological history of the Alps and Carpathians from Krezsek and Bally, 2006; Kuhlemann, 2007.
6 European river histories. Danube: Gábris and Nádor, 2007; Olariu et al., 2018. Rhone: Sissingh, 1997; Gargani, 2004. Rhine: Kuhlemann, 2007; Preusser, 2008. Meuse: van Balen et al., 2000. Vistula: Brud, 2004. Po: Amorosi et al., 2008; Ghielmi et al., 2013. Ebro: Garcia-Castellanos et al., 2003. Rivers of northern France and southern England: Antoine et al., 2003; Bridgland et al., 2006; Westaway et al., 2006.
7 Hötzl, 1996; Hill and Polyak, 2014.
8 Berendsen and Stouthamer, 2001; Pierik et al., 2017.
9 Müller, W. 'My heart, do you now recognize' Wigmore, R. (1988) *Schubert: The Complete Song Texts*. New York: Schirmer Books, translation of Poem 7.
10 Information about Wilhelm Müller and Franz Schubert from Baumann, 1981; Brown, 1983.
11 Medieval Rhine floods from Herget and Meurs, 2010; Herget et al., 2015.
12 Information about Murchison from Geikie, 1875; Stafford, 1989. His Russian travels from his unpublished diaries, reproduced in Collie and Diemer, 2004.
13 Stafford, 1989.
14 Humboldt's Russian travels from Botting, 1973, pp. 238–252; Naumann, 2007.
15 Murchison, R. 'thrown up in vertical beds … at least one thousand of my lazy officers' (1845). Collie, M. and Diemer, J. (eds)(2004) *Murchison's Wanderings in Russia*. Nottingham: British Geological Survey Occasional Publication No. **2**, 238 and 226.
16 Murchison, 1845.
17 Geological history of the Volga, Don, Dneiper, and Dneister from Obedientova, 1977, p. 40; Brunet et al., 2003; Popov et al., 2006; Vincent et al., 2010, 2013.
18 Geological history of the Tigris and Euphrates from Potter and Hamblin, 2006; Demir et al., 2007; Nicoll, 2010.
19 Eide et al., 2017.
20 Napoleon at the Berezina River from Nicolson, 1985.
21 Napoleon 'the lives of a million men'. Nicolson, N. (1985) *Napoleon: 1812*. London: Weidenfeld and Nicolson, p. 177.

Chapter 8: The Reversing Rivers of South America
1 Amazon exploration and indigenous population from Smith, 1879, pp. 18–33; De Bruhl, 2010; Clement et al., 2015.
2 Padre Simon 'he was growing morose'. Quoted in Smith, H.H. (1879) *Brazil, the Amazons and the Coast*. New York: Charles Scribner's Sons, 22.
3 Smith, 1879, pp. 24–25, 168–170.
4 Clement, 1999.
5 Amazon River data from Hovius, 1998; Archer, 2005; Wohl, 2007.
6 Tectonic history of South America from Willis, 1944; Jordan et al., 1983; Horton, 2018.
7 Humboldt's travels from Botting, 1973; Wulf, 2014.
8 Winemiller et al., 2008.

9 Crist, 1932.

10 Humboldt, A. 'I was ever aware'; 'So you're interested in botany?' Quoted in Botting, D. (1973) *Humboldt and the Cosmos*. London: Sphere Books, pp. 176 and 179.

11 Raby, 2001.

12 South American rivers from Potter, 1997; Cox, 1989; Hoorn et al., 1995, 2017; Diaz de Gamero, 1996; Théveniaut and Freyssinet, 2002; Albert et al., 2006; Potter and Hamblin, 2006; Figueiredo et al., 2009; Latrubesse et al., 2010; Shephard et al., 2010; Makaske et al., 2012.

13 Leake, 2011, p. 201-207.

Chapter 9: Canyons and Cataracts in North America

1 Ann Weiler Walka 'In my room above a crowded, wintry street' (2000) *Waterlines*. Flagstaff: Red Lake Books.

2 Expedition of Lt. Ives from Ives, 1861.

3 Ives, Lt. J.C. 'The region last explored' (1861) *Report upon the Colorado River of the West*. Washington: Government Printing Office, 36th Congress, 1st Session, House Executive Document No. **90**, Part 1, p. 110.

4 Ives, 'The extent and magnitude of the system', ibid, p. 109.

5 Newberry, J.S. 'the most splendid exposure … lacks a single requisite' (his italics) (1861) *Report upon the Colorado River of the West*. Washington: Government Printing Office, 36th Congress, 1st Session, House Executive Document No. **90**, Part 1, pp. 42, 45–46.

6 John Wesley Powell and his Grand Canyon journey from Powell, 1895; Stegner, 1954; Worster, 2001.

7 Worster, D. 'Wes discovered rivers' (2001) *A River Running West: The Life of John Wesley Powell*. New York: Oxford University Press, p. 76.

8 Stegner, 1954, p. 195.

9 Stegner, 1954, p. 95.

10 Stegner, 1954, p. 147.

11 Information about G.K. Gilbert from Pyne, 2007.

12 Orme, 2007.

13 Dutton, C. 'At the foot of this palisade' (1882) *Tertiary history of the Grand Cañon district, with atlas*. Washington: US Government Printing Office, pp. 146, 149, 150.

14 William Morris Davis and his Colorado Plateau work from Davis, 1900, 1909; Chorley et al., 1973.

15 Davis, W.M. 'precocious young valley' (1909) The lessons of the Colorado Canyon. *Bulletin of the American Geographical Society* **41**, 345–354.

16 Colorado River from Ranney, 2012, 2014; additional information from Dexter, 2009; Sears, 2013; Karlstrom et al., 2014; Hill and Polyak, 2014; Darling and Whipple, 2015; Howard et al., 2015.

17 Lee, K. 'Pore Colly Raddy' (2006) *Glen Canyon Betrayed: A Senuous Elegy*. Flagstaff: Fretwater Press, p. 69.

18 Tectonic history of North America from Silver and Smith, 1983; Grand et al., 1997; Monger and Price, 2002.

19 Lee, 2006, p. 157.

20 Dutton, C. 'No doubt the question' (1882), ibid., p. 260.

21 Information about Mark Twain from Twain, 1883, 1924; Ferguson, 1965; Kaplan, 1974.

22 Moody et al., 2003.

23 Twain, M. 'It is good to begin life poor' (1924) *The Autobiography of Mark Twain*, p. 25.

24 Twain, 1924, 'No – afraid you would', ibid., p. 11.

25 Twain, M. 'piloting becomes another matter' (1883), ibid.

26 Twain, 1924, 'Old Silurian Invertebrates', ibid., pp. 82–83.

27 Kaplan, 1974, p. 39.

28 Twain, M. 'It was a monstrous big river' (1884) *Adventures of Huckleberry Finn*.

29 Ferguson, 1965, pp. 54–55.

30 Kaplan, 1974, p. 15.

31 Mississippi River from Mann and Thomas, 1968; Potter, 1978; Archer and Greb, 1995; Galloway et al., 2011; Blum and Pecha, 2014.

32 Obermeier et al., 1990.

33 River histories of North America from Heller et al., 2003; Mack et al., 2006; Galloway et al., 2011; Blum and Pecha, 2014.

34 Columbia and Snake rivers from Smith, 1988; Beranek et al., 2006; Reidel and Tolan, 2013.

35 Pierson, 1985.

36 Rivers of Atlantic Canada and offshore Maine from Pe-Piper and MacKay, 2006; Falcon-Lang et al., 2007, 2016; Li et al., 2012; Zhang et al., 2014; Chavez et al., 2019.

37 Rivers of the northeastern USA from Cleaves, 1989; Poag et al., 1990; Miller et al., 2013; Naeser et al., 2016.

38 Clark, William 'the waters of the Kansas'. From notebooks of the Lewis and Clark expedition, archived at the American Philosophical Society, Philadelphia.

Chapter 10: A Canadian Amazon

1 Human occupation of the Cypress Hills from Hildebrandt and Hubner, 2007.

2 Geology of Cypress Hills and nearby areas from Storer, 1978; Leckie and Cheel, 1989.

3 Zaslow, 1975, p. 155.

4 Information in this section from Ottawa Evening Journal, May 1895.

5 O'Brien, 1982.

6 Information about Robert Bell from Ami, 1927; Zaslow, 1975.

7 Ami, H.M. 'We had two buckboards' (1927) Memorial of Robert Bell. *Bulletin of the Geological Society of America* **38**, p. 26.

8 Bell, R. 'A Pre-glacial River' (1895) A great pre-glacial river in northern Canada. *Scottish Geographical Magazine* **11**, p. 368.

9 Labrador Sea geology from McMillan, 1973; Duk-Rodkin and Hughes, 1994; Hiscott, 1984; Duk-Rodin et al., 2004; Jauer and Budkewitsch, 2010.

10 McMillan and Duk-Rodkin, 1995.

11 Blum and Pecha, 2014.

12 Norris, 1993.

Chapter 11: Frozen Out: Northern Rivers Sculpted by Ice

1 Klondike gold rush and geology from Adney, 1900; Berton, 1972; Lowey, 2006.

2 Lowey, 2006.

3 Yukon, Mackenzie, and Fraser rivers from Duk-Rodkin and Hughes, 1994; Duk-Rodkin et al., 2001, 2004; Froese et al., 2005; Lowey, 2006; Andrews et al., 2012; Hidy et al., 2013.

4 Shugar et al., 2017.

5 Information about Louis Agassiz from Lurie, 1960; Imbrie and Imbrie, 1979.

6 Imbrie and Imbrie, 1979, p. 25.

7 Geikie, 1875, pp. 303–310.

8 Lurie, E. 'God's great plough' (1960) *Louis Agassiz: A Life in Science*. Chicago: The University of Chicago Press, p. 98.

9 Geikie, A. 'Agassiz gave us a great field-day … If you have not been frost-bitten' (1875) *Life of Sir Roderick I. Murchison, v. 1*. London: John Murray, pp. 307 and 308.

10 Lurie, 1960, p. 120.

11 Maclaren, C. 'If we suppose the region' (1842) The glacial theory of Professor Agassiz of Neuchâtel. *American Journal of Science* **42**, p. 365.

12 Peltier and Fairbanks, 2006.

13 Information about James Croll from Imbrie and Imbrie, 1979, pp. 77–96.

14 Information about Milutin Milanković from Imbrie and Imbrie, 1979, pp. 97–122; Greene, 2015.

15 Imbrie, J. and Imbrie, K.P. 'the little room seemed like the nightquarters' (1979) *Ice Ages: Solving the Mystery*. Short Hills: Enslow Publishers, 102.

16 Hays et al., 1976.

17 Pleistocene Ice Age in Eurasia and North America from Dyke et al., 2002; Sejrup et al., 2005; Hidy et al., 2013; Niessen et al., 2013; Stokes et al., 2016; Cordier et al., 2017.

18 Engineering the Falls: The Corps Role at St. Anthony Falls. 'in the rocky clefts'. Retrieved on 2 January 2012 from http://www.mvp.usace.army.mil/history/engineering/

19 Winchell, 1878.

20 Fisher, 2003.

21 O'Farrell, 1995.

22 Gunn, 1982.

23 Geology of Ireland, rivers, Ice Age history, and Irish elk from Jukes, 1862; Farrington, 1965; Mitchell, 1980; Stuart et al., 2004; Sejrup et al., 2005; Simms and Coxon, 2017.

24 Information about W.B. Yeats from Hone, 1962.

Chapter 12: Megafloods and Noah's Ark

1 Information about J Harlen Bretz from Soennichsen, 2008.

2 Bretz, 1923, with additions from Bretz, 1925.

3 Bretz, J.H. 'a feeling of amazement' (1923) The channeled scablands of the Columbia Plateau. *Journal of Geology* **31**, p. 621.

4 Gould, 1978; Baker, 1995.

5 Hakon Wadell's work from Wadell, 1920; Bretz, 1964; Soennichsen, 2008, p. 178.

6 Soennichsen, J. 'If you know you are right' (2008) *Bretz's Flood*. Seattle: Sasquatch Books, p. 195.

7 Pardee, 1942; Baker, 1995.

8 Weis and Newman, 1989.

9 Soennichsen, J. 'We are all now catastrophists' (2008), ibid, p. 231.

10 Scabland floods from Waitt, 1985; O'Connor and Baker, 1992; Zuffa et al., 2000; Clague et al., 2003; Waitt et al., 2019.

11 Bruchac, 2005.

12 Glacial lakes and megafloods in North America and Eurasia from Carling, 1996; Larson and Schaetzl, 2001; Mangerud et al., 2001; Clarke et al., 2004; Herget, 2005; Lewis and Teller, 2007.

13 Russell and Knudsen, 2002; Snorrason et al., 2002; roadside signs in Iceland.

14 Waitt, 2002; Baynes et al., 2015.

15 Middle Eastern excavations from Ryan and Pitman, 1998; Ryan, 2007.

16 The Epic of Gilgamesh, 'the uproar of mankind'.

17 Mudie et al., 2004; Hiscott et al., 2007; Turney and Brown, 2007.

18 Chepalyga, 2007; Badertscher et al., 2011.

Chapter 13: Rivers Drowned by the Sea

1 Mecham, G. 'I could not but fancy' (1855) *Traveling report in: Recent Arctic expeditions in search of Sir John Franklin and the crews of H.M.S. 'Erebus' and 'Terror'*. London: G.E. Eyre and W. Spottiswoode, p. 502.

2 Beaufort Formation from Fyles, 1990; Rybczynski, 2008; Rybczynski et al., 2013; Braschi, 2015.

3 Bernard Pelletier, Obituary; Frisch, 2015.

4 Ancestral Arctic rivers from Pelletier, 1966; Bornhold et al., 1976.

5 Sea-level changes from Dott, 1992.

6 Stanley and Warne, 1994.

7 Gustaaf Molengraaff and his work from Molengraaff, 1921; Brouwer, 1941–2.

8 Sunda Shelf from Pelejero et al., 1999; Clode and O'Brien, 2001; Hanebuth et al., 2011; Alqahtani et al. 2017.

9 Raby, 2001.

10 Nunn and Reid, 2016.

11 Amber information from Poinar, 1992.

12 Mythology of amber from Pliny the Elder.

13 Eridanos River from Gibbard et al., 1988; Overeem et al., 2001; Cohen et al., 2014; Gibbard and Lewin, 2016.

14 Rose and Rosenbaum, 1993; Barton et al., 2004.

15 Information about Bill King from Shotton, 1963a, b.

16 Barton, P., Doyle, P. and Vandewalle, J. 'I said, You have done a lot of exploring' (2004) *Beneath Flanders Fields, The Tunnellers' War 1914–1918*. Montreal: McGill-Queen's University Press, p. 74.

17 Strahan, 1917; King, 1919, 1921; Doyle, 1998; Barton et al., 2004.

18 Information about Dudley Stamp and his work from Stamp, 1923, 1927; Kimble, 1967.

19 Shotton, 1963b.

20 King, 1954; Smith, 1989.

21 Summerfield and Whalley, 1980; Jones, 1999; Newell, 2014.

22 Channel River from Smith, 1985, 1989; Gupta et al., 2007, 2017; Toucanne et al., 2010; Westaway and Bridgland, 2010; Winsemann et al., 2016; García-Moreno et al., 2019.

23 Terraces on Channel River tributaries from Antoine et al., 2003; Westaway et al., 2006; Lewin and Gibbard, 2010.

24 Bridgland, 2013.

25 Thomas, E. 'alive with the dead' (1909) *Richard Jefferies*. London: Faber and Faber.

26 Jefferies, R. 'I was not more than eighteen' (1883) *The Story of My Heart*. London: Longmans, Green and Company.

27 Jefferies, 1879, p. 17.

28 *Pheasant*, by Bryan Hanlon.

29 Thomas, E. 'children whose naked feet' (1909), ibid.

Chapter 14: From Stone Age Streams to River Civilizations

1 Hominin history and environments in East Africa from Ashley et al., 2010, 2014.

2 Tagore, 1931, p. 38.

3 Semaw et al., 2003; Brunet et al., 2005; Harmand et al., 2015; Humphrey and Stringer, 2018.

4 Schlebusch et al., 2017.

5 Twain, 1924, p. 279.

6 Geomorphic work of non-human animals from Darwin, 1883; McCarthy et al., 1998; Jones and Gustason, 2006; Rybcynski, 2008; Butler and Malanson, 2005; Butler, 2006; Johnson et al., 2010.

7 Roebroeks and Villa, 2011; Gibling, 2018.

8 Head, 1989; Turney et al., 2001.

9 Jericho excavations from Kenyon, 1957; Hillel, 1994, pp. 152–158.

10 Bar-Matthews et al., 1997.

11 Tengberg, 2012.

12 Barkai and Liran, 2008.

13 Weisdorf, 2005.

14 Bar-Yosef, 1998; Balter, 2010.

15 Helmer et al., 2005; Wilkinson, 20015; Larson et al., 2007; Tarolli et al., 2014; Arranz-Otaegui et al., 2016; Stephens et al., 2019.

16 Dietrich et al., 2012.

17 Moore et al., 1984; Denham et al., 2003; Rosen, 2008; Walter and Merritts, 2008; Brown et al., 2013; Gibling, 2018.

18 Mays, 2008.

19 Lightfoot, 1996; Weinberger et al., 2008.

20 Frumkin and Shimron, 2006.

21 Hillel, 1994, p. 64-72; Erickson-Gini, 2012.

22 Hillel, 1994, p. 42.

23 Weiss et al., 1993.

24 Hillel, 1994, p. 45.

25 McCully, 1996, p. 13; Wilkinson, 2003.

26 Hillel, 1994, pp. 55–56, 95.

27 Morozova, 2005.

28 Hillel, D. 'I constrained the mighty river to flow … I dug a canal' (1994) *Rivers of Eden*. New York: Oxford University Press, p. 53.

29 Hillel, D. 'I dug a canal' (1994), ibid, p. 301.

30 Hillel, 1994, pp. 41, 47–50, 58–59; Jacobsen and Adams, 1958.

31 Weiss et al., 1993; Weiss and Bradley, 2001; Staubwasser and Weiss, 2006.

32 Herodotus, Book 1, parts 189–191.

33 The Epic of Gilgamesh, 'Here in the city man dies oppressed'.

34 Hillel, 1994, pp. 111–113.

35 Quaternary climate of North Africa and the Nile from deMenocal et al., 2000; Macklin et al., 2013.

36 Butzer, 1976, p. 20-21, 89; Hillel, 1994, pp. 58–59.

37 Mays, 2008.

38 Herodotus, Book 2, part 100.

39 Hillel, D. 'enabled Egyptian farmers' (1994), ibid., p. 59.

40 Agriculture in South and Central America from Clement et al., 2015; Goman et al., 2005; Dillehay et al., 2007; Ertsen and van der Spek, 2009; Ranere et al., 2009; Haas et al., 2013; Hilbert et al., 2017; Watling et al., 2018.

41 McNamee, 1994, pp. 27–29; Reisner, 1993, p. 256; Huckleberry, 1999.

42 Rice genetics and cultivation from Fuller et al., 2010; Liu et al., 2007; Molina et al., 2011; Huang et al., 2012; Hilbert et al., 2017.

43 Rosen, 2008.

44 Christian, 2000.

45 Rossel et al., 2008; Outram et al., 2009; Qiu et al., 2015; Almathen et al., 2016.

46 Kierstein et al., 2004.

47 Sharma et al., 1980.

48 Harvey and Fuller, 2005; Gibling et al., 2008.

49 Roffet-Salque et al., 2015.

50 Tacitus, pp. 62, 66, 114–115; Grimal, 1983, pp. 68–72; Caracuta and Fiorentino, 2011.

51 Bono and Boni, 1996; Leo and Tallini, 2007; Mays, 2008; Beard, 2016.

52 Marchetti, 2002; Bruno et al., 2013.

53 Kamash, 2012.

Chapter 15: The Lost Saraswati River of the Indian Subcontinent

1 Information about *Rigveda* from Staal, 1996, 2008; Danino, 2010.

2 Information about Max Müller from Chaudhuri, 1974.

3 Griffith, R.T.H. 'The music of the frogs' (1889) *The Hymns of the Rgveda*. Book, hymn and verse: 7, 103, 2–10.

4 Danino, 2010, pp. 13–15.

5 Oldham, C.F., 1874, 1893; Oldham, R.D., 1886.

6 Oldham, C.F. 'such meteorological conditions' (1893) The Saraswati and the lost river of the Indian Desert. *Journal of the Royal Asiatic Society of Great Britain and Ireland* **34**, 52.

7 Wilhelmy, 1969.

8 Marshall, 1924. Other information about the Indus Civilization from Wheeler, 1962; Lahiri, 2000; Albinia, 2010, pp. 245, 253–4; Danino, 2010.

9 Danino, 2010, pp. 91–93, 26–140.

10 Wheeler, 1962, p. 6; Lambrick, 1967; Lahiri, 2000, p. 19; Danino, 2010, pp. 182–183.

11 Tandon et al., 1997; Srivastava et al., 2001.

12 Tod, 1829, pp. 19–20.

13 Luni River from Enzel et al., 1999; Jain et al., 2005.

14 Recent research on the Saraswati in Yash Pal et al., 1980; Saini et al., 2009; Clift et al., 2012; Sinha et al., 2013; Singh et al., 2016; Singh et al., 2017; Khan and Sinha, 2019; Singh and Sinha, 2019.

15 Singh et al., 2017.

16 Kipling, 1901, p. 151.

17 Albinia, 2010, p. 221; Danino, 2010, p. 50.

18 Griffith, R.T.H. 'whose limitless unbroken flood' (1889), ibid. Book, hymn and verse 6.61, parts of 8–14; 1.3.12.

Chapter 16: Confucian Engineers on the Yellow River of China

1 Lamouroux, 1998.

2 Harris, 2008.

3 Ball, 2017, p. 20.

4 Ball, 2017, p. 60–61.

5 Viollet, 2007, p. 236; Ball, 2017, pp. 90–91.

6 Viollet, P.-L. 'The river broke through at Huzi' (2007) *Water Engineering in Ancient Civilizations: 5,000 Years of History*. Boca Rotan: Taylor and Francis Group, p. 243.

7 Lamouroux, 1998, p. 560.

8 Will, 2006, pp. 383–343.

9 Ball, 2017, pp. 43–95.

10 Gascoigne, 2003, p. 59.

11 Viollet, 2007, p. 231.

12 Xu, 2003.

13 Elvin, M. and Ts'ui-jung, L. 'When the sediments are scoured away' (1998) The influence of the Yellow River on Hangzhou Bay since A.D. 1000. In Elvin, M. and Ts'ui-jung, L. (eds) *Sediments of Time, Environment and Society in Chinese History*. Cambridge: Cambridge University Press, 401.

14 Dodgen, R.A. 'Year after year' (2001) *Controlling the Dragon: Confucian Engineers and the Yellow River in Late Imperial China*. Honolulu: University of Hawai'i Press, p. 137.

15 Information in the following sections from Dodgen, 2001.

16 Dodgen, R.A. 'The people work on the river' (2001), ibid., p. 112.

17 Ball, 2017, pp. 23, 186.

18 Dodgen, R.A. 'I stood still for a time' (2001), ibid., pp. 67–68.

19 Huang et al., 2004, 2006; Rosen, 2008; Zhuang and Kidder, 2014.

20 Chu et al., 2002; Ge et al., 2013.

21 Economy, 2010, p. 44.

22 Crook, D. and Elvin, M. 'The mountains whose sides (2013) Bureaucratic control of irrigation and labour in late-imperial China: the uses of administrative cartography in the Miju catchment, Yunnan. *Water History* **5**, p. 303.

23 Crook, D. and Elvin, M. 'People's survival is not to be treated' (2013), ibid., p. 292.

24 Crook, D. and Elvin, M. 'I've been clearing the river out' (2013), ibid., p. 304.

25 Ball, 2017, pp. 189, 210.

26 Lary, 2001; Dutch, 2009; Muscolini, 2011.

27 Thomson, 1992.

28 Grypma, 2012.

29 Morris, 1970, p. 196.

30 Davis, A.R. 'A man who has feelings' (1971) *Tu Fu*. New York: Twayne Publishers, p. 53.

31 Hinton, D. 'Azure heaven pinched between Wu Mountains' (1996) *The Selected Poems of Li Po*. New York: New Directions Publishing Corporation, Starting Up Three Gorges, 100.

Chapter 17: Dead and Wounded Rivers

1 Leopold, 1949, p. 200.

2 Information about John Muir from Muir, 1916; Worster, 2008.

3 Muir, J. 'I allow people to find out' (1916) *A Thousand-Mile Walk to the Gulf*. Boston: Houghton Mifflin.

4 Muir, J. 'honorable representatives of the great saurians' (1916), ibid.

5 Worster, D. 'I will touch naked God' (2008) *A Passion for Nature; The Life of John Muir*. Oxford: Oxford University Press, p. 215.

6 Muir, J. 'No doubt these trees' (1920) *Nature Writings*. Save the Redwoods.

7 Worster, 2008, p. 233.

8 Muir, J. 'Dam Hetch-Hetchy! … these temple destroyers' (1908) Hetch Hetchy Valley.
9 McCully, 1996, p. 13; Lenders et al., 2016.
10 Walter and Merritts, 2008.
11 Freese, 2003, pp. 103–128.
12 Twain, 1924, pp. 109–110; Ferguson, 1965, pp. 66–77; Kaplan, 1974, pp. 22, 43.
13 Macy, 2010.
14 California gold mining and Gilbert's assessment from Gilbert, 1917; James, 1999; Pyne, 2007, pp. 203–254.
15 Macy, 2010, pp. 21–25, 391–393.
16 Reisner, 1993; Farmer, 1999, pp. 131–2.
17 Reisner, 1993, p. 136.
18 Statistical information from Wong et al., 2007; Lehner et al., 2011; Jaramillo and Destouni, 2015; Canadian Dam Association; International Commission on Large Dams; International Rivers; Headwater Economics, 2016; National Dam Inventory, 2016; WWF, 2018.
19 Middleton, 2012, p. 96.
20 Fearnside, 2004; Deemer et al., 2016.
21 Vörösmarty et al., 2010; Schleiss et al., 2016; Creed et al., 2017.
22 Kondolf et al., 2014; Winemiller et al., 2016; Latrubesse et al., 2017.
23 Macy, 2010, p. 11.
24 McCully, 1996, pp. 17, 26.
25 McCully, 1996, pp. 237–238.
26 Xiangtang Hong 'Have you crossed the Yellow River? (2009) *Performing the Yellow River Cantata.* Unpublished PhD thesis, University of Illinois at Urbana-Champaign.
27 Pietz, 2015, p. 173.
28 Guang Weiran 'The old world must be changed!'(1958), Sanmenxia de he chang.
She kan **7**, pp. 68–73.
29 Information about Huang Wanli from Shapiro, 2001, pp. 23, 48–65.
30 Shui Fu 'Some mistakes can be remedied' (1998) A profile of dams in China. In Dai Qing (ed.) *The River Dragon Has Come!* Armonk: M.E. Sharpe, p. 21.
31 Ran et al., 2013; Chu, 2014.
32 Chu, 2014.
33 Xu, 2004.
34 Leopold, 1998; Yang et al., 2011; Zhang and Lou, 2011; Dai and Liu, 2013.
35 Xie, 2003; Turvey et al., 2010.
36 Barnett et al., 2015; Yang et al., 2015.
37 Dai Qing 'I think that 90 per cent … If the Three Gorges could speak' (1998) *The River Dragon Has Come!* Armonk: M.E. Sharpe, xxix, xvii.
38 O'Brien, 1982.
39 Examples in this section from Achterberg et al., 1999; Cordos et al., 2003; Stradling and Stradling, 2008; Kelly et al., 2010; Ruyters et al., 2011; Chartin et al., 2013; Gulin et al., 2013; Liu and Hua-Jiang, 2013; Dew et al., 2015; Petticrew et al., 2015; Sansone et al., 1996; Segura et al., 2016; Thompson et al., 2020.
40 Wong et al., 2007.

Chapter 18: Collapsing and Closing Dams
1 Jarrett and Costa, 1985; Blair, 1987.
2 Reisner, 1993, pp. 379–410.
3 Yi Si, 1998.
4 Shui Fu, 1998; Ball, 2017, p. 229.
5 McCully, 1996, p. 115.
6 Flooding behind the High Aswan Dam from Dafalla, 1975.
7 Michaels, A. 'The silt, like the river water' (2009) *The Winter Vault.* London: Bloomsbury, p. 34.

8 Michaels, A. 'The Great Temple had been carved out' (2009), ibid. pp. 139 and 332.
9 Nile Delta from Simpson et al., 1990; Hillel, 1994, pp. 128–131;Stanley, 1996; Stanley and Warne, 1998; Nicholls et al., 2007.
10 Clements, 1959.
11 Clements, F. 'I've never known a river like this' (1959) *Kariba.* London: Methuen & Company, p. 167.
12 Hughes, D.M. 'the greatest environmental upset' … 'No single project' (2006) Whites and water: How Euro-Africans made nature at Kariba Dam. *Journal of Southern African Studies* **32**, 825–826, 837.
13 Powell, J.W. 'Sometimes the rocks are overhanging' (1895) *Canyons of the Colorado* (later published unabridged as *The Exploration of the Colorado River and its Canyons*).
14 Farmer, 1999, pp.129–135.
15 Powell, 2008.
16 Abbey, E. 'obliterating from a sandstone wall' (1968) *Desert Solitaire.* New York: Touchstone, p. 38.
17 Stegner, 1969, p. 129; Lee, 2006, p. 179.
18 Farmer, 1999, p. 180.
19 Abbey, E. 'The beavers had to go and build' (1968), ibid., pp. 151–152.
20 Abbey, E. 'alone in the silence' (1968), ibid., p. 191.
21 Abbey, E. 'Dear old God' (1975) *The Monkey Wrench Gang.* New York: HarperCollins Publishers, p. 33.
22 Abbey, E. 'where great waterfalls … will breathe a metaphorical sigh of relief' (1968), ibid., pp. 127 and 267.
23 Farmer, 1999, p. XIII; Myers, 2013; Lustgarten, 2016; United States Bureau of Reclamation, 2017.
24 Farmer, 1999, pp. XVII, 111, 117, 162.
25 Stegner, W. 'In gaining the lovely' (1969) *The Sound of Mountain Water.* Garden City: Doubleday & Company, p. 128.
26 Leslie, 2005.

Chapter 19: Between the Dams: An Elegy for the Saskatchewan River
1 Lewis, 2004.
2 Smith et al., 1998; Smith et al., 2014.
3 Morozova and Smith, 2000.
4 The Partners for the Saskatchewan River Basin, 2009.
5 Herriot, 2000, p. 101.
6 Schindler and Donahue, 2006.
7 COSEWIC, 2006.
8 Smith et al., 2016.
9 World Wildlife Fund, 2009.
10 Kulchyski et al., 2007.
11 Kulchyski et al., 2007, pp. 30–32; Loney, 1987.
12 Nelson and Churchill systems from Chodkiewicz and Brown, 1999; Kulchyski et al., 2007; Neckoway, 2007.
13 Carlson, 2008.
14 Carlson, H.M. 'When you talk about money' (2008) *Home is the Hunter: The James Bay Cree and Their Land.* Vancouver: UBC Press, p. 218.
15 Carlson, H.M. 'This is what I want you to understand'(2008), ibid., p. 249.
16 Olson, 1961.
17 Dostoevsky, F. 'Love all God's creation' (1880) *The Brothers Karamazov.*

Chapter 20: Without Spoiling the Land: Rivers and Agriculture
1 Charles Hartt and Herbert Smith in Brazil from Hartt, 1874a, 1874b, 1879; Smith, 1879; Brice and Figueirôa, 2003; Woods et al., 2009.

2 Hartt, C.F. 'These pests disappear at night' (1874b) Preliminary report of the Morgan Expeditions, 1870-71. Report of a reconnaissance of the Lower Tapajos. *Bulletin of the Cornell University (Science)* **1**, 18.

3 Roosevelt et al., 1996.

4 Smith, 1879, pp. 136–144, 168–172, 380; Woods et al., 2009.

5 Anthropogenic dark soils from Glaser et al., 2001; Woods et al., 2009; Schmidt et al., 2014; Clement et al., 2015; Watling et al., 2018.

6 Leopold, A. 'the standard paradox of the twentieth century' (1938) Engineering and conservation. In S.L. Flader and J.B. Callicott (1991) *The River of the Mother of God*. Madison: The University of Wisconsin Press, pp. 249–254.

7 Powell, 1878.

8 Smith, 1946.

9 Stegner, W. 'I tell you gentlemen' (1954), ibid., p. 343.

10 Stegner, W. 'It was the West itself that beat him' (1954), ibid., p. 338.

11 Libecap and Hansen, 2002.

12 Soennichsen, 2008, p. 175.

13 Worster, 1979, pp. 87, 97.

14 Reisner, M. 'set out to help the small farmers' (1993), ibid., pp. 485–486.

15 Information in this section from Leopold, 1949; Meine, 1988.

16 Herriot, 2000, p. 36.

17 Dust Bowl information from Hoyt, 1938; Worster, 1979; Cook et al., 2009.

18 Leopold, A. 'the river was nowhere and everywhere' (1949) *A Sand County Almanac, and Sketches Here and There*. London: Oxford University Press, p. 142.

19 McCully, P. 'The unregulated Colorado' (1996) *Silenced Rivers*. London: Zed Books, p. 46.

20 Luecke et al., 1999.

21 Leopold, 1938, pp. 249–250.

22 Meine, C. 'The building of a power dam' (1988) *Aldo Leopold. His Life and Work*. Madison: The University of Wisconsin Press, pp. 512–513.

23 Meine, C. 'My own impression' (1988), ibid., p. 482.

24 Meine, 1988, p. 370.

25 Leopold, A. 'We and they had found a common home' (1949), ibid., p. 148.

26 Contreras et al., 2012; Goñi, 2018.

27 Wright and Wimberley, 2013.

28 Wohl, 2005.

29 Micklin, 1988; Millennium Ecosystem Assessment, 2005; Albinia, 2010; Pearce, 2017.

30 Meine, C. 'The greatest fortune I can wish you' (1988), ibid., p. 302.

Chapter 21: London's Buried Rivers

1 Hill, 1955; Mackintosh, 1955.

2 London rivers and their geological and human history from Barton, 1962; Clayton, 2000; Bridgland, 2003; Bridgland et al., 2006; Halliday, 2009; Clements, 2010.

3 Pepys, 1660.

4 Halliday, 2009, p. 40, and history of London's sanitation in chapters 1–2.

5 Botting, 1973, pp. 227–228.

6 Morley, H. (1854), September 30. Sick Body, Sick Brain 'It comes home to the minister of state' *Household Words*, **X**, 148. Retrieved on 10 November 2017 from http://www.djo.org.uk/household-words/

7 Halliday, S. 'If there be sufficient authority' (2009) *The Great Stink of London*. Stroud: The History Press, ix - xi.

8 Weintraub, 1993, p. 373; Tomalin, 2011, pp. 227–228.

9 Halliday, 2009, chapters 4, 7 and 8.

10 Schneer, 2005, pp. 149–160.

11 Barton, 1962, p. 125.

12 Clayton, 2000, p. 99.

13 Plaque in basement of Grays Antique Market, 1–7 Davies Mews, Mayfair, London.

14 Clayton, A. 'Well, you can take it from me' (2000) *Subterranean City*. London: Historical Publications, 33–34.

15 Death's Doors, 1854.

16 Dickens, C. 'glistening water tinted with the light' (1836–7) *The Pickwick Papers*.

Chapter 22: Restored Rivers

1 Information in this section from The Laxá Farmers, 2013.

2 The Saga of the People of Reykjadal and of Killer-Skuta 'absolutely worthless' In V. Hreinsson (ed.) (1997) *The Complete Sagas of the Icelanders*. Reykjavik: Leifur Eiríksson Publishing, 302.

3 Information in this section from Cronin and Kennedy, 1997; Forbes, 2004.

4 Cronin, J. and Kennedy, R.F.Jr. 'But they never asked about triplicates' (1997) *The Riverkeepers*. New York: Scribner, p. 32.

5 Cronin, J. and Kennedy, R.F.Jr. 'If Patton didn't get me killed' (1997), ibid., p. 47.

6 Fielding, 2011.

7 Forbes, L.C. 'You can't look at those young faces' (2004) Pete Seeger on environmental advocacy, organizing, and education in the Hudson River Valley. *Organization & Environment* **17**, 513–522.

8 Brown et al., 2013; O'Connor et al., 2015; Tonra et al., 2015; Headwater Economics, 2016; Schiermeier, 2018.

9 Leopold, 1938, p. 250.

10 Barnett and Pierce, 2008; Lustgarten, 2016.

11 Davison, 2017; New Zealand declares a river a person, 2017.

12 Safi, 2017.

13 Atwood, 2008, p. 180.

14 Herriot, T. 'Although we have done much' (2000) *River in a Dry Land: A Prairie Passage*. Toronto: Stoddart Publishing Company, p. 13.

Epilogue

1 Fletcher, 1967, pp. 140, 146, 217–220.

2 Jefferies, R. 'It is eternity now' (1883) *The Story of My Heart*

INDEX